M000159191

MIND BEYOND
BRAIN

MIND BEYOND BRAIN

BUDDHISM, SCIENCE, AND THE PARANORMAL

DAVID E. PRESTI

WITH
BRUCE GREYSON, EDWARD F. KELLY,
EMILY WILLIAMS KELLY,
AND JIM B. TUCKER

Columbia University Press   New York

Columbia University Press
Publishers Since 1893
New York   Chichester, West Sussex
cup.columbia.edu
Copyright © 2018 Columbia University Press
All rights reserved

Library of Congress Cataloging-in-Publication Data
Names: Presti, David E., editor.
Title: Mind beyond Brain : deepening the dialogue between science
and Buddhism / [editors] David E. Presti with Bruce Greyson,
Edward F. Kelly,
Emily Williams Kelly, and Jim B. Tucker.
Description: New York : Columbia University Press, [2018] |
Includes bibliographical references and index.
Identifiers: LCCN 2018010141 (print) | LCCN 2018029059 (ebook) |
ISBN 9780231548397 (electronic) | ISBN 9780231189569 (cloth : alk. paper)
Subjects: LCSH: Parapsychology—Religious aspects—Buddhism. |
Occultism—Religious aspects—Buddhism. | Religion and science. |
Buddhism.
Classification: LCC BQ4570.P75 (ebook) |
LCC BQ4570.P75 M56 2018 (print) |
DDC 294.3/365—dc23
LC record available at https://lccn.loc.gov/2018010141

Columbia University Press books are printed on permanent
and durable acid-free paper.
Printed in the United States of America

Cover design: Guerrilla Design
Cover illustration: Mandala painting by Anil Thapa,
Lumbini Buddhist Art Gallery, Berkeley, CA

The spectacular progress of the physical sciences since the seventeenth century was made possible by the exclusion of the mental from their purview. To say that there is more to reality than physics can account for is not a piece of mysticism: it is an acknowledgment that we are nowhere near a theory of everything, and that science will have to expand to accommodate facts of a kind fundamentally different from those that physics is designed to explain. It should not disturb us that this may have radical consequences. . . . Materialism remains a widespread view, but science does not progress by tailoring the data to fit a prevailing theory.

Thomas Nagel

Science cannot solve the ultimate mystery of nature. And that is because, in the last analysis, we ourselves are part of nature and therefore part of the mystery that we are trying to solve.

Max Planck

*There does not exist anything
That is not dependently arisen.
Therefore there does not exist anything
That is not empty.*

Nāgārjuna

CONTENTS

FOREWORD

GESHE TENZIN WANGYAL RINPOCHE

When I first met a few years ago with the researchers who contributed to this book, I was excited to learn about their work. These different areas of study—near-death experiences, memories of past lives, apparitions, *siddhis*—reflect a principle that Tibetan Buddhists have explored for centuries. We believe that mind is not just conditioned by the physical brain and body, or even by the properties of time and space. Mind is also beyond all these. Even after the dissolution of the physical body at the time of death, Tibetan Buddhists believe that mind persists and is reborn continuously into new forms.

Mainstream science has been slow to acknowledge the dimensions of study described in this book. Given the relevance of these topics to the contemporary encounter between Buddhism and science, in 2010, I invited these researchers to participate in a Buddhism-science conference at Serenity Ridge, Virginia—the retreat center where I focus my own worldwide efforts in research, study, and teaching (Ligmincha International). This setting permitted engaging in open, lively dialogue, and the presentations at that conference were the inspiration for this book.

This gathering was a realization of some of my own long-time interests. From age ten, I was trained as a Bön monk in Menri Monastery near Dolanji, India. I entered the Dialectic School at Menri, and after finishing twelve years of study there toward my *geshe* degree, I traveled to the West and discovered here a completely different world. I was excited to explore on an informal basis everything from psychology to neuroscience to quantum physics. In particular, I have always been interested in how ancient spiritual practices can bring about not only profound shifts in consciousness but also physical health benefits.

Around 1993 I met some clinicians at the University of Texas M.D. Anderson Cancer Center, and we opened discussions around the relationship between meditation practice and pain management. This eventually led to projects in which we observed the mental and physical effects of Tibetan practices—of the body (such as the Tibetan yogas of *tsa lung* and *trul khor*), of breath (various breathing and sound-healing exercises), and of mind (including meditation practices involving imagery and visualization)—on the relief of suffering in cancer patients.[1] Ancient Tibetan tantric and *dzogchen* texts tell of a wide variety of benefits from these practices, based on centuries of observation in meditation.[2] I believe that practices of the body, breath, and mind will one day be acknowledged by conventional medicine as potent treatments for physical disease, and will be prescribed for specific bodily symptoms as well as to improve the health of the psyche.

The researchers in this book have been observing and cataloging unusual phenomena that are difficult or impossible to explain based on current scientific theory. Near-death experiences bring up the question of how we can reconcile a brain that appears not to be functioning with a mind functioning better than ever. Cases of the reincarnation type suggest that memories and emotions can carry over from one life to the next. Apparitions associated with death indicate that the process of dying may be far more interesting and revealing than simply the cessation of the physical functioning of the body. Investigation of extraordinary powers or attainments (*siddhis*) such as telepathy, clairvoyance, and psychokinesis presents yet another perspective on an expanded view of mind and reality.

It is well accepted in the Tibetan traditions that body, energy, and mind are inseparably related, and that the mind itself has extraordinary power not only to ease physical suffering and prevent and heal sickness, but also to dissolve the entire physical dimension. For example, through practices of wind (*prana, chi, lung*), one is able to deconstruct the energetic blockages within one's bodily tissues until those blockages dissolve and one enters a spacious state that is free of discord and pain. Enlightenment itself is said to come about when one deconstructs the dualistic sense of ego identity to the point that one attains a pure state of consciousness (*kunzhi*, the base of all consciousness) that is beyond all conceptuality and even physicality. In the Tibetan *dzogchen* tradition, there are records of people whose physical bodies at the time of death dissolved entirely or almost entirely into light, with the exception of hair, fingernails, and toenails. This highest attainment of *dzogchen* practice is known as the body of light, or rainbow body (*gya lü*).

On a more practical and achievable level, when presenting these practices to students and cancer patients I have focused on deconstructing the sense of ego identity so that one identifies, essentially, as a totally new person with a much healthier, more open, and caring relationship to pain. I have seen over and over how the warmth of loving kindness toward one's mental, emotional, or physical pain can ease and even eradicate pain and sickness.

Some of the effects of these practices are so profound and subtle that they are difficult to measure. And some of the research presented in this book faces similar challenges of measurement. But I truly believe the time will soon come when these areas of research will be well accepted. The contributors in this book are courageous to openly explore their research in the face of skepticism by many of their peers, and I think it is important they continue. What we can learn from these investigations may be applied to something much more important than trying to read someone's mind, talk to spirits, or determine one's previous lives. Investigation into mind and its relationship with our physical world, which has been conducted since ancient times by countless wisdom traditions, can not only expand upon the current limits of our science, but can offer something very concrete in meeting the needs of modern society. This book serves as an excellent vehicle in bringing very important material more fully into the contemporary encounter between Buddhism and science. It is good to see this happen.

PROLOGUE: DEEPENING
THE DIALOGUE

DAVID E. PRESTI AND EDWARD F. KELLY

The explanatory scenario of contemporary science is one of awesome power and beauty. This grand human enterprise has described many natural phenomena in stunning detail—from subatomic particles to the diversity of living organisms, from neural signaling in brains and bodies to exploding supernovae deep in space. Within this framework, much can be calculated, computed, predicted, engineered, and constructed. Applications extend from automobiles and rockets, to integrated circuits and communications technology, to agriculture and medicine. Most of us are immersed all day every day in products derived from our scientific understanding of the world.

Yet, many mysteries remain. One of the premier academic forums for the publication of scientific research is the journal *Science*. In celebration of its 125th anniversary in 2005, the journal assembled a list of 125 outstanding unanswered scientific questions.[1] The first two questions on this list concerned the nature of the universe and the nature of mind or consciousness.

The nature-of-the-universe question was formulated in terms of the concepts of dark matter and dark energy, used to describe the fact that the standard model of Big Bang cosmology is unable to explain the matter and energy character of what is estimated to be 96 percent of the observable universe.[2] This question could be reformulated as: what is the nature of the physical universe, where did it come from, and where is it going? Or, more broadly: what is the nature of reality?

The consciousness question was formulated in terms of understanding how brain processes relate to conscious experience.[3] This question is essentially asking: what is the nature of the human mind? Whence comes our capacity to have thoughts, feelings, perceptions, and awareness—the qualities of conscious experience? Or, more broadly: who are we, really, and how are we related to the rest of the physical universe?

DIALOGUE BETWEEN BUDDHISM AND SCIENCE

This book deals with those topics: the nature of reality and the nature of mind. It is offered as a contribution to the evolving dialogue between the traditions of science and of Buddhism, a dialogue that seeks to bring together the best parts of these complementary approaches to the study of mind.[4] Western science (which has now become global science) has emphasized third-person, objective measures, in which the system being investigated is observed, probed, and measured from the "outside," using procedures that can be replicated by other scientists. Buddhism, on the other hand, has developed highly refined first-person observation of subjective experience, using procedures that can also be replicated.

The interaction of Buddhism with science is an important aspect of Buddhism's encounter with modernity. It may be dated as beginning in the mid-nineteenth century and it has many facets.[5] Nonetheless, the dialogue between Buddhism and science achieved new levels of promise and prominence when, in the 1980s, His Holiness the 14th Dalai Lama[6] began a series of conversations with scientists from the West. Initially taking place between the Dalai Lama, psychologists, cognitive scientists, and neuroscientists,[7] the dialogue soon expanded to include physicists, educators, and environmental scientists.[8] New directions of research in the neuroscience of meditation and in the psychology of emotion have emerged from these interactions.[9]

The Dalai Lama has been interested in science since he was a young boy. He has spoken and written of his childhood fascination with gadgets and of his wish to understand how things work, and he has long appreciated that both science and religion are associated with understanding the nature of who we are and how we are related to the rest of what we call reality. He has also recognized that the complementary approaches to observation and analysis employed by Buddhism and modern science may each have things of value to offer. He summarizes his perspective on the current conversation in his book *The Universe in a Single Atom: The Convergence of Science and Spirituality*.[10]

With an eye toward increasing knowledge about contemporary science among those who are already expert in Buddhist philosophy and practice, the Dalai Lama has initiated or inspired several programs designed to introduce science education into Tibetan Buddhist monastic communities in India.[11] This is a remarkable achievement and the Dalai Lama has indicated that his contribution to increasing dialogue between Buddhism and science is, in his opinion, among his most important accomplishments.[12] In 2012, the Dalai Lama received the Templeton Prize, an award recognizing exceptional contribution to humankind's spiritual dimension. In announcing this award, specific mention was made of the Dalai Lama's fostering of "connections between the investigative traditions of science and Buddhism as a way to better understand and advance what both disciplines might offer the world."[13]

THE PRESENT BOOK IN THE CONTEXT OF THE ONGOING DIALOGUE

This book grew from a day-long symposium that took place in October 2010 at the Serenity Ridge Retreat Center of the Ligmincha Institute near Charlottesville, Virginia. Ligmincha Institute was founded and is guided by Geshe Tenzin Wangyal Rinpoche, a master teacher in the Tibetan Bön tradition. While the Bön tradition traces its lineage to the shamanic culture of pre-Buddhist Tibet, there is much similarity between Bön as currently practiced and the various schools of contemporary Tibetan Buddhism.[14] All are concerned with analysis of the nature of mind, and with the development of ethical behavior, compassion, and equanimity.

Tenzin Wangyal was born in India and received monastic education there. In 1991 he came to the United States and in 1992 founded the Ligmincha Institute in Virginia. In 2009 he became acquainted with the very unusual but also highly relevant work of a small research unit in the psychiatry department of the University of Virginia Medical School in Charlottesville—the Division of Perceptual Studies (DOPS).

DOPS was founded in 1967 by Ian Stevenson (1918–2007). At the time, Stevenson was Chair of the Department of Psychiatry at the University of Virginia Medical School. During his more than half-century at the University of Virginia, Stevenson carried out pioneering work in the empirical investigation of the mind-body connection, focusing on phenomena suggesting that contemporary scientific hypotheses concerning the nature of mind, and the mind's relation to matter, may be seriously incomplete. Stevenson authored a dozen books and more than two hundred scientific publications related to this work.[15]

Stevenson was a meticulous scientist and conservative in his interpretation of data. At the same time, he was a true visionary, and courageous in his study of phenomena that lie outside the current scientific mainstream. Under Stevenson's leadership, DOPS became the largest and longest-running university-based group in the United States devoted exclusively to empirical investigation of phenomena that are not easily encompassed by the current paradigm in the biophysical sciences for understanding the mind. This paradigm maintains that consciousness is entirely generated by, or emerges from, the physical processes of the brain and body, and can in no way be more than that (chapter 1).

One line of research conducted by DOPS involves "near-death experiences" (NDEs)—powerful and transformative events sometimes experienced by individuals who come close to death (chapter 2). Many NDEs take place under extreme physiological conditions such as cardiac arrest or general anesthesia, conditions for which contemporary neuroscience would predict that conscious experience ought to be severely diminished or abolished altogether. Some NDEs also include reports that experiencers are able to accurately perceive their physical situation from a perspective outside their own body, a phenomenon that is impossible to explain within the current neurobiological descriptions of mind-brain connection.

Another major program of research at DOPS focuses on "cases of the reincarnation type" (CORT), in which small children, generally ages two to five years, begin to speak and act as though they are remembering persons, places, and events from

another, usually recent, previous life (chapter 3). These memories can sometimes be investigated, and occasionally the stated
details prove verifiable to a considerable and startling degree.
Cases have included, for example, extremely unusual birthmarks
or birth defects corresponding to injuries or fatal wounds associated with the remembered previous life.

Another arena of work at DOPS involves studies of mediumship, crisis apparitions, and deathbed visions—phenomena
suggesting that the connection between mind and body mani-
fests in unusual ways around circumstances involving death or
other physical crises (chapter 4).

Finally, DOPS houses a state-of-the art neuroimaging facility devoted to laboratory-based studies of "psi" phenomena,[16]
such as telepathy and precognition, as well as altered states
of consciousness that appear to be conducive to these sorts of
events (chapter 5).

In general, within the academic community of scientists and
scholars, there is little awareness or appreciation of the amount
and quality of data supporting phenomena such as NDEs,
CORT, mediumship, apparitions, and psi. However, there *is*
a great deal of empirical evidence suggesting that these phenomena do happen. Knowing this, Tenzin Wangyal invited the
principal DOPS researchers to come to Serenity Ridge for a
one-day meeting to present and discuss their various research
activities.

The topics presented in this book are deeply embedded not
only within Buddhism, but also within many other religious and
spiritual traditions. An in-depth discussion of these topics might
have already taken place within the Buddhism-science dialogue as it has evolved so far, but it has not. Very early on in the
encounters between the Dalai Lama and groups of scientists—
in 1989 during the second of what are called the Mind & Life

conferences—the Dalai Lama initiated a discussion of rebirth and the possibility of memories from prior lives. He indicated that there is credible evidence for what appear to be memories from other lives and presented an example with which he was familiar. From the record of this discussion, the neuroscientists with whom he was having the conversation appeared to not appreciate that such phenomena could be investigated using the methods of science, nor were they informed about attempts to study such phenomena within the world of Western science.[17] While some of these topics arose again in subsequent meetings, there has yet to be an in-depth discussion.[18]

The phenomena discussed in this book present a serious challenge to the dominant hypothesis in contemporary neuroscience—that mind derives entirely from the local physical functioning of the brain and body in some relatively straightforward, though still largely unknown, manner. It is clear that such a hypothesis cannot account for things like veridical reports of memories from other lives, or out-of-body experiences in which a person is able to accurately describe perceptions from locations well outside his or her body. It is also clear that if the standard neuroscientific model of mind is correct, then things central to many spiritual traditions simply are not possible.

In an excellent essay characterizing the current dialogue between Buddhism and science, Thupten Jinpa, scholar of Buddhism and long-time translator of the Dalai Lama, writes that:

> the more metaphysical aspects of the two traditions—the concepts of rebirth, karma, and the possibility of full enlightenment of Buddhism; and physicalism, reductionism, and the causal enclosure principle on the part of the scientific worldview—are left bracketed.[19]

Here "bracketed" means these concepts are not part of the discourse—they are off the table as possible topics of discussion. The presumed reason for this is that being "more metaphysical" they are situated outside the scope of the scientific method. We disagree, as will become clear. Jinpa goes on to say that:

> [an] objective in this critical engagement with science has to do with responding to the perceived challenges being posed by the scientific worldview to key Buddhist concepts. What is the status of the concept of rebirth? What does the closure principle, which states that the only causes that operate are material, imply for the Buddhist theory of karmic causation? How does Buddhism respond to the widely assumed regulative principle in cognitive science that mind equals brain and that all mental states are, in the final analysis, merely brain states?[20]

Excellent questions. In a companion essay, scholar of Buddhism Donald Lopez writes that:

> . . . karma, rebirth, and the possibility of full enlightenment are among the most important foundations of Buddhist thought and practice. Physicalism, reductionism, and the causal closure principle also are highly important, especially in neuroscience. These are precisely the topics that must be unbracketed and confronted in any discussion of Buddhism and science. It is also among these topics that the most intractable disagreements likely lie.[21]

We heartily agree. It is time to bring these concepts onto the table for discussion and to shake up the conversation a bit. If the dialogue is to truly flourish, then, says scholar of Buddhism

José Cabezón, it must be understood that Buddhism and science are "systems that can both support and challenge each other at a variety of different levels—no monopolies, no holds barred!"[22]

As will become clear to readers of this book, we believe that the science of mind is poised for the entrance of revolutionary new ideas—and perhaps on the verge of a paradigm shift such as only occurs very infrequently in the history of science. Delving into places of extraordinary challenge and "intractable disagreement" can surely contribute to catalyzing this process. Our book is offered in this spirit—in hopes of alerting readers from *both* sides of the dialogue to scientific research that focuses strongly on the deepest and most challenging aspects of the conversation.

SCIENCE AND SPIRITUALITY

Questions about the nature of mind can easily evoke associations to *spirit* and *soul*, and the many different definitions, connotations, and emotional reactions people may have related to these terms.[23] Investigation—even solid scientific investigation—of who we are and how we are related to the rest of the universe can bring one close to what is often considered the domain of religion, and apparently outside the domain of science. This proximity can be disturbing—to individuals in either camp.

Opinions about dialogue concerning science, religion, and spirituality vary widely. At least in the Western world of European and American culture, there is often a perception of tension between science and religion. Galileo's persecution by the Roman Catholic Church in the 1600s for his defense of the idea that the earth moves around the sun stands as an oft-quoted and overly simplified example. And these days, prominent scientists and other authors have written best-selling books dismissing religious belief as toxic and delusional. Others have spoken of a middle path, describing science and religion as non-overlapping domains of practice.[24] However, none of this does justice to the far more nuanced and complex relationship science and religion have had over the centuries, and all the way up to the present time. Even the very categories of what we now call *science* and what we now call *religion* are, it can be argued, relatively recent distinctions.[25]

Questions regarding the ultimate nature of reality and of mind *are* of interest both to modern science and to religious and spiritual traditions. And it may be that the evolving dialogue between science and spirituality in its various forms represents *the* major intellectual challenge of our present era.

MIND BEYOND BRAIN

1

SCIENTIFIC REVOLUTION AND THE MIND–MATTER RELATION

Science: a process of collecting and organizing knowledge about the world, the universe, nature. It derives from the Latin word *scire*, to know. The scientific method has come to mean a process of collecting data through observation and experimentation, guided by the formulation of questions and hypotheses, and further deepened by theoretical frameworks of "explanation" that endeavor to connect observations together in ever richer networks. The result is enhanced "understanding."

Revolution: from the Latin *revolvere*, to turn away. A revolution is a turning from one way of envisioning things to another, dramatically different, way. In the history of science over the last five hundred years, there have occurred only a very small number of revolutionary turning points before and after which the world was described very differently.[1]

Mind and *Consciousness:* by *mind* we will mean the collection of possible mental states—thoughts, feelings, and perceptions. These are irreducibly subjective states of experience, part of *what it is like to be* you. By *consciousness* we mean awareness or sentience. Awareness of what? Awareness of these subjective mental states. There are also mental states out of awareness, nonconscious mental states—the unconscious aspects of mind. In Western cultures, one of the tasks of psychotherapy is to bring unconscious mental material into conscious awareness, where it may be subjected to analysis and change. In Eastern cultures, this is one of the tasks of yoga—to become ever more aware of and conversant with all aspects of mind. By these definitions, mind and consciousness share much in common, and indeed the two terms are often used interchangeably.[2]

These definitions of mind and consciousness are circumscribed and personal; they do not go beyond a person's own experience. However, some uses of these terms reference something more expansive—something transpersonal, transcendental, nonlocal, or universal. Is consciousness a personal experience, a process that only makes sense subjectively? Or is it a presence in the cosmos, perhaps even existing outside of or beyond space and time? This can create problems right out of the gate that may interfere with the development of a conversation. Thus, we begin with the more circumscribed, local, personal definition, to get the conversation started, and see where this takes us, returning to these distinctions later in the book. If a science of consciousness is to flourish, precise definition and use of terms will help to prevent confusion. Additional terms can be introduced if and when necessary.[3]

Matter: the stuff of the physical universe; the substance of which physical objects are composed.

Metaphysics: the stage upon which our description of the world is played out, the framework within which the results of our scientific inquiry are interpreted. The metaphysical frame for contemporary science is called *physicalism*, sometimes referred to as *physical materialism*. Physicalism posits that all of what we call reality is conceived as constructed in some way from matter and its interactions as described by the mathematical laws of physics. While it is the opinion of many practicing scientists that their work is conducted without any preconceived notions or bias, the physicalist metaphysical assumption is generally implicit in the way science is conducted and may impose significant constraints on our capacity to ponder the nature of reality and the nature of mind.

THE AWESOME EXPLANATORY FRAME OF CONTEMPORARY PHYSICAL SCIENCE

The explanatory framework of contemporary science extends from the submicroscopic domain of particles making up the internal structure of atoms to the vastness of the cosmos. How did this come to be? In 1543, Nicolaus Copernicus (1473–1543) published *On the Revolutions of the Heavenly Spheres*, his magnum opus describing a sun-centered (heliocentric) interpretation of the movements of the sun and planets. A few years later, Johannes Kepler (1571–1630) described the movements of the planets around the sun in terms of simple mathematical relations. Galileo Galilei (1564–1642) provided further observations of celestial movements supporting a heliocentric interpretation of the solar system and took steps toward developing mathematical descriptions of the movement of terrestrial objects—such as a swinging pendulum or an object falling from a tower. Not long after, Isaac Newton (1643–1727) set forth a comprehensive description of motion that applied equally to celestial and terrestrial phenomena. An apple falling from a tree, the rising and falling of oceanic tides, the trajectory of a cannonball, the movement of the Earth and other planets around the sun, the movement of moons around a planet, the trajectories of comets crossing the heavens from deep in space—all were described by the same mathematically elegant laws of motion. An impressive unity of explanation had been achieved.

A key aspect of this scientific revolution was a focus on describing a physical world as it appears to us, treating the world as if it exists independently of our perception of it. That mind was left out of the equation was famously articulated by René Descartes (1596–1650), who described the domain of science to be that of the material world (*res extensa*), including the physical body. Mental phenomena (*res cogitans*)—thoughts, conscious awareness, subjective experience, the seat of the human soul— were the domain of the spirit, falling outside the purview of investigation by physical science and within the purview of religion. This split of mind from the world (and from the body) was done in part to protect the project of science from being influenced by religious institutions. At the time Descartes (a Frenchman living in the Netherlands) was writing on this subject, he was well aware that Galileo was suffering censure at the hands of the Catholic Inquisition in Italy.

Over the next two centuries, the mathematical foundations put in place by Newton were extended to apply to a wide variety of natural phenomena—light, electricity, magnetism, gaseous pressure, temperature, and other thermodynamic properties among them. This description of the so-called physical properties of the world is called "classical physics." In this context, a "physical property" is understood as an aspect of reality describable in terms of mathematical properties ascribed to coordinates in space and time. Mass, energy, electric charge, and momentum are examples of physical properties. This historical trajectory—beginning with Copernicus and extending into the twentieth century—was one of revolutionary reconceptualization of nature.[4]

At the beginning of the twentieth century, Albert Einstein (1879–1955) expanded the explanatory framework of physics with his formulations of the special and general theories of relativity. Einstein's contributions were revolutionary. Although still operating within the metaphysical framework of physicalism, special and general relativity radically transformed our conceptions of space, time, mass, and energy. Under appropriate conditions, mass and energy are now known to be interconvertible, according to $E = mc^2$. And space and time are no longer simply the static backdrop upon which events take place, but rather are malleable and transformed by the relative motion of observers, and by the presence of mass and energy.

Only a few years after Einstein's relativistic revolution, there was another major revolution in physics. In the first decades of the twentieth century, it was found that the known physical laws failed to account for the behavior of electrons and protons interacting to form atoms. In order to deal with the experimental observations of the properties of atoms and molecules, it was necessary to develop an entirely new physical theory, with new physical concepts and new dynamical equations. This was quantum mechanics—the collective work of a number of physicists all engaged in deriving a mathematical theory that would account for the properties of atoms. Among the authors of this truly revolutionary undertaking were Max Planck, Albert Einstein, Niels Bohr, Louis de Broglie, Werner Heisenberg, Erwin Schrödinger, Paul Dirac, Max Born, Wolfgang Pauli, and others. This was a profound revolution, and nearly a century later there remains a lack of consensus regarding what quantum mechanics implies about the nature of reality—what exactly might be the metaphysical implications of the theory.[5]

It is clear that our current physical theories of nature have powerful explanatory utility. Quantum mechanics together with special relativity yield a relativistic quantum field theory that along with the so-called standard model of elementary particle physics provides a very successful description of the measurable properties of matter. This includes the structure of atoms and molecules, the strong and weak nuclear interactions describing the internal structure of atomic nuclei and the interactions of various fundamental particles, and the properties of visible light and other forms of electromagnetic energy. General relativity provides a description of gravitational interactions and the large-scale structure of the cosmos. Physicists agree that there remains an incompleteness that might be addressed by somehow combining quantum mechanics and general relativity into a unified theory of all matter and energy. Multidimensional string theories are an attempt to accomplish this. Some physicists hold that we may be but a small conceptual step away from a grand unified theory of everything. Witness these words from the final paragraph of a recent book by physicist Stephen Hawking (1942–2018)[6]:

> The fact that we human beings—who are ourselves mere collections of fundamental particles of nature—have been able to come this close to an understanding of the laws governing us and our universe is a great triumph. . . . If the theory [specifically, m-theory, a current version of 11-dimensional string theory] is confirmed by observation, it will be the successful conclusion of a search going back more than 3,000 years. We will have found the grand design.[7]

MEANWHILE, IN BIOLOGY . . .

Within a physicalist metaphysics, concepts from physics and chemistry provide the foundation for the biological sciences. And as the sciences of physics and chemistry have expanded in explanatory scope and elegance over the last several centuries, so also has biology. Long a descriptive science lacking unified coherence, biology was revolutionized in the mid-nineteenth century by the idea that biological diversity comes about through continual processes of variation and selection. Working from physical evidence that the Earth possesses great age and from many observations of plants and animals under a variety of conditions, Charles Darwin (1809–1882) and Alfred Russel Wallace (1823–1913) put forth the concepts that have come to be called biological evolution.[8] This basis for appreciating a profound interconnectivity of all life on Earth was articulated at length in Darwin's justifiably famous book, *On the Origin of Species by Means of Natural Selection*, published in 1859. The immense age of the Earth provided ample time during which variation in the properties of organisms could take place (the nature of the underlying variation was at that time unknown) and selection of particular organismal traits would occur as a result of environmental conditions—either the natural environment or at the hands of humans, as in the case of domesticated plants and animals.

The revolution of evolution took on additional momentum in the twentieth century when, with rapidly expanding knowledge of biochemistry, the quest for the molecular basis of life took center stage. Of particular interest was the great mystery of heredity: how was the information needed to construct a living organism stored within a cell and transmitted from one generation to the next? In a lecture delivered in 1932, Niels Bohr (1885–1962)—one of the physicists who had a few years earlier contributed to the founding of quantum mechanics—speculated that perhaps the laws of physics and chemistry, as then currently understood, might be inadequate for describing the fundamental nature of living organisms.[9] Bohr saw this, however, as a reason not for pessimism but rather for its opposite: just as the analysis of matter at the atomic level had precipitated the discovery of the new physical laws of quantum mechanics, so perhaps the pursuit of the molecular basis of life might also entail the discovery of completely new physical principles. What could be more exciting than that?

Bohr's lecture was inspirational to Max Delbrück (1906–1981), a young physicist who heard it at the time, and who went on to become one of the pioneers of the yet-to-be-founded scientific field of molecular biology. Delbrück—who was one of the first to consider that genetic information in organisms was somehow encoded within the structure of large molecules—went on to develop experimental techniques to investigate the molecular basis of heredity using bacteria and viruses.[10] The ideas of Bohr and Delbrück were further elaborated in a widely read little book entitled *What Is Life?*, published in 1944 by another of the founding physicists of quantum mechanics, Erwin Schrödinger (1887–1961). This book set forth the idea that new physical principles—so-called laws of nature—might be discovered in the quest to understand the molecular basis of life. This provided inspiration to a number of scientists in the early days of molecular biology.[11]

The scientific revolution begun by Darwin achieved a culmination of sorts with the molecular description of deoxyribonucleic acid (DNA) in 1953,[12] the description of the genetic code within the following decade, and more recently, the sequencing of the human genome and the genomes of an increasing number of other organisms. The source of genetic variation, unknown and mysterious in the nineteenth century, is now understood as molecular changes in the nucleic acid sequences of an organism's genome. At least thus far, the speculation that new laws of nature might be discovered in the quest to understand life has not panned out. Increasingly sophisticated descriptions of the molecular basis of heredity—and other chemical and physical processes underlying life at the cellular and molecular levels—have been developed without the need to invoke any radically different explanatory frameworks or new physical forces. Indeed, contemporary molecular biology has seen triumph after triumph in the application of a physicalist framework leading to a richer understanding of the nature of life.

Within this framework, living organisms are understood as particular structures composed of atoms and molecules, organized so as to utilize energy in the maintenance of stability, information storage, and replication.[13] The diversity of life is considered to have developed over billions of years according to the principles of biological evolution, and we humans and our capacities are understood as part of this grand explanatory scenario. Although some big mysteries remain—perhaps the most notable of which are the emergence of life from nonliving matter and the emergence of sentience within life—there is confidence among most scientists that a complete explanation of life and mind will be found within something similar to our present metaphysical framework and corpus of physical laws.

From the Big Bang origin of the physical universe to the magnificent diversity of life in Earth's biosphere, we have an explanatory scientific narrative of stunning beauty. And we humans love a good story.

HOW DOES MIND FIT IN?

Within this framework, it is assumed that mind and consciousness have developed according to principles of biological evolution. That is, whatever the capacity for mental experience may be, it has evolved in intimate concert with body physiology and is an essential part of how our body engages with the world around it.[14] It certainly *could* be the case that a capacity for conscious awareness gives us (and other creatures who may possess such awareness) an enhanced ability to analyze and integrate information—and to share that information through social cognition and language—facilitating survival and reproduction in the complex environment of our planet's biosphere.

A wealth of empirical data suggests that our capacity for mental experience is intimately related to the material substance of our body and in particular of our brain. A pioneer of the study of mind within modern Western science, William James (1842–1910),[15] stated it clearly in *The Principles of Psychology* (1890):

> If the nervous communication be cut off between the brain and other parts, the experiences of those other parts are non-existent for the mind. The eye is blind, the ear deaf, the hand insensible and motionless. And conversely, if the brain be injured, consciousness is abolished or altered, even although every other organ in the body be ready to play its normal part.[16]

Thus, damage to the brain is associated with specific changes in mental functioning. In addition, certain chemical substances—psychoactive drugs such as stimulants, sedatives, psychedelics, and so forth—that enter the brain are associated with specific effects on mental experience. And neural activity of the brain—as measured by imaging technologies such as electroencephalography (EEG), functional magnetic resonance imaging (fMRI), and positron emission tomography (PET)—is correlated in specific ways with perceptions, emotions, and other mental processes.[17]

That there is an intimate relationship between mind and body is unquestioned. However, just how this relationship is manifest—how it is that the physical processes taking place among the atoms, molecules, and cells of the brain and body are related to the subjective experiences of mind and consciousness—is a deep mystery. Many who reflect upon the mind-body relationship have concluded that it appears to be a very difficult problem. Indeed, how subjective experience is related to brain and body physiology has been termed the "hard problem" of consciousness research. It is said that there is an "explanatory gap" between mind (subjective experience) and brain, body, and more generally, matter (physical stuff).[18]

Within a strictly physicalist metaphysics—where there is a necessity that mind and consciousness be explained in terms of the properties of matter—this mind-body relationship may always be a problem. For there is a difference of category between the mental and the physical—consciousness is irreducibly subjective and experiential, and thus very different from physical stuff. That is precisely what makes the mind-body problem hard—some would say impossible.

Various approaches to the mind-body problem have been put forth over the centuries. Here is an abbreviated summary of terms used in Western philosophy. There is *physicalism*, about which we have been speaking, wherein mind must be derivable from the physical properties of matter, and it is matter that is really real. There is *dualism* (or *substance dualism*), in which there are two separate irreducible domains—mind (mental stuff/ experience) and matter (physical stuff)—that come together and interact within, for example, the human body. There is *idealism*, where what is really real is the mental domain and the physical world is derived from, and only exists as per, our experience. There is *panpsychism*, in which mind and consciousness are intrinsic and fundamental in some way to the fabric of reality. There is *neutral monism*, where mind and matter are considered dual complementary aspects of an underlying neutral, beyond concept, reality. These terms, and more, are used in philosophy to describe different metaphysical frameworks for understanding the mind-matter connection.[19]

One approach to the mind-body problem from the perspective of contemporary biology and physicalist metaphysics is to recognize that life provides a separation between inner and outer. All living organisms have an interior-exterior boundary. Single-celled microorganisms (prokaryotic bacteria and eukaryotic protists) are in sensory contact with their environment—detecting the presence of nutrient chemicals and sunlight. Thus, even relatively simple life forms can be imagined to have some sort of inner perspective on sensory information.

As sensory processes became more elaborate over the course of evolutionary history, that inner perspective became correspondingly more nuanced—for example, animal vision presumably provides a more elaborate view of the surrounding environment than the light-sensing processes in phototactic bacteria. A major task of animal brains is the analysis of information collected by sensory organs. Sensory information is collected, integrated, and utilized by the nervous system to inform and orchestrate the movement of the animal's body through the world. Thus, perhaps the first inklings of inner experience came along with the evolutionary development of sufficiently complex nervous systems. In this view, the capacity for mind and consciousness is inherent in life,[20] and sophisticated sentience goes along with nervous systems that conduct a sufficiently sophisticated analysis of sensory information.[21]

Perhaps conscious awareness provides a richer representation of sensory information, allowing for more elaborate possibilities of interaction with the world around us. Perhaps in some yet inexplicable manner it allows for new kinds of causal effects on brain processes and thereby on behavior. And perhaps all this results in evolutionarily adaptive value above and beyond the absence of awareness. Contrasting with this view is the hypothesis that awareness is epiphenomenal, merely noticing or noting what has already, and inexorably, been set in place by the physical processes operating within the body, the nervous system, the brain—like a radio or television announcer creating the play-by-play description at a sporting event.

FUTURE TRAJECTORIES IN THE INVESTIGATION OF MIND AND CONSCIOUSNESS

Consciousness: from the Latin *con*, meaning with, and *scire*, knowledge or knowing, the same Latin root as science. Consciousness is our most salient experience. Through consciousness we are *aware* that we know things about the world—we *experience* knowing. All organisms, whether they have consciousness or not, know things—be it as simple as day from night, light from dark, up from down, or the presence or absence of nutrients. However, without consciousness, they will not be *aware* of the knowing. Via consciousness we notice patterns and regularities and develop explanatory frameworks in which we "understand" the patterns and regularities as aspects of a reified, external, objective, "real" world—a world that is assumed to exist independently of our awareness and within which we have evolved through a process of biological evolution governed by physical laws. While the existence of a "real" world external to us is extrapolated and assumed, we only come to know this assumed objective world via our experience. *There is an inextricable enfolding of mind and world.* That said, how might we continue to move forward with a scientific investigation of mind and consciousness?

Science is grounded in *empiricism*—from the Greek *empeirikos*, meaning experience, or experiment. The term originally referred to a school of ancient physicians who based their medical practice on observations and data from their own experience, rather than on dogmatic theory. Let us stick with the scientific method of empirical observation and experiment, while maintaining a broadened stance where nothing is left out, including even questioning the metaphysical framework—honoring that all we know ultimately comes via our experience, and explicitly acknowledging the inextricable enfolding of mind and world. Here are several suggested trajectories:

Expanded development of biophysical science. We can honor the excellent track record of physical science in describing the universe by continuing to do everything we are doing to better understand the structure and function of the physical world, including brains and bodies. Keep dissecting brains, recording neural activity, mapping neural connections; investigating genetics and the regulation of gene transcription; and probing the microscopic interior structure of cells. Continue to push empirical biophysical science as far as possible. These sorts of projects already enjoy widespread support within orthodox mainstream science. Certainly, many discoveries will be made along this trajectory, hopefully leading to improved understanding of health and disease and relief of suffering for many beings.

Fundamental physics and consciousness. As we endeavor to understand mind and consciousness in terms of biophysical science, we should allow for the contribution of fundamental aspects of physics in ways not presently appreciated. Quantum physics appears to suggest this already. How so?

Much of our experience of the world is interpretable in terms of classical (Newtonian) physics. Objects are observed to move along well-defined trajectories and interact in ways akin to colliding billiard balls. However, as mentioned earlier in this chapter, quantum mechanics has replaced classical physics as the most comprehensive known description of the behavior of matter and energy in the universe. And quantum physics indicates that the behavior of matter and energy at the microscopic level is very weird indeed. For example, an elementary particle, such as an electron, seemingly exists in many states and places simultaneously—a situation described by a quantum-mechanical wave function, a mathematical construct assigning probabilities to various alternative possibilities.

However, we experience reality as actualities, not potentialities—electrons, protons, atoms, and molecules, for example, having actual locations in space and time. How this transition from potential to actual comes about is called the "measurement problem" in quantum physics. It is vigorously debated among physicists interested in the foundations of quantum physics – what physics has to say about the nature of reality and our ability, through experiment and observation, to acquire knowledge about reality. There are many opinions among physicists, some strongly held, but there is no consensus save for an appreciation that a deep mystery exists here. There are strong reasons to believe that the participatory experiential role of the observer/experimenter/scientist *cannot* be removed from understanding what quantum physics says about the world.[22] This could be an important clue.

In a different direction, another topic of investigation at the frontier of modern physics is the notion that there exist additional dimensions beyond our usual three dimensions of space and one dimension of time.[23] Additional dimensions, if they are to be granted a kind of physical reality, may offer the basis for an explanatory framework for all kinds of otherwise inexplicable phenomena, including some or even all of the things described in the present book.[24] This is interesting to at least consider. Perhaps all this suggests that the next really big thing in physics will incorporate a central role for consciousness within the context of physical science. Just as we look to experiments in high-energy particle physics and to observations in cosmology and astrophysics to catalyze new ideas in fundamental physics, so might it be that investigations of the nature of mind and life will lead to unexpected new ideas in physics. This was the speculation of Niels Bohr, Max Delbrück, and Erwin Schrödinger almost a century ago. Perhaps the time has come.

Refined analysis of mental experience. William James suggested in his writings that a science of mind be based upon a multifaceted empirical approach: study of behavior (psychology), study of the biological underpinnings of behavior (neurobiology), and study of the mental phenomena themselves (introspection and phenomenology). The first two approaches have been extensively developed in the century since James. However, the direct investigation of mental experience has not yet seen the same level of development within the scientific enterprise.[25]

James argued in *The Principles of Psychology* that empirical study of mental experience was essential to a science of mind: *"Introspective Observation is what we have to rely on first and foremost and always.* The word introspection need hardly be defined— it means, of course, the looking into our own minds and reporting what we there discover."[26] But careful introspective observation requires a sustained focus of attention internally—on the contents of one's own mind—and sustained focusing of attention does not come easily. James again spoke to this: "the faculty of bringing back a wandering attention, over and over again, is the very root of judgment, character, and will. . . . An education which should improve this faculty would be *the* education *par excellence*. But it is easier to define this ideal than to give practical directions for bringing it about."[27]

Sophisticated methods to train attention, coupled with introspective observation and analysis of mind, have been explored for millennia by contemplative traditions. This exploration is perhaps most evident in Buddhism. Scholar and teacher of Buddhism Alan Wallace has written extensively about the valuable ways Buddhist contemplative practices might contribute to an expanded Western science of mind:[28]

> The mind-body problem, which remains in the domain of philosophical speculation, calls for an unprecedented expansion of the scientific method. Integrating scientific and contemplative modes of inquiry in the exploration of the mind and its origins may enable us to finally solve it. This will not occur as long as our starting assumptions about the mind are materialistic and our research methods observe only physical behavior and neural correlates of mental states and processes. In all branches of natural science, the most revolutionary insights are gained by directly and meticulously observing the phenomena under investigation. Observation of the mind itself is the strength of the contemplative traditions of the world, and the union of contemplative and scientific methods may yield a true contemplative science that revolutionizes our understanding.[29]

This trajectory is served by continued development of a collaborative conversation between science and Buddhism. From this may also come deeper appreciation of the unique qualities of a very different worldview, one grounded in a science of mental experience. Psychologist Eleanor Rosch, who was a participant in the very first Mind & Life dialogue with the Dalai Lama in 1987, put it this way:

> The least recognized time bomb of the 20th century may be contact between the Asian meditation traditions and Western Culture. At their best, these traditions offer a portal into a radically new (lived) understanding of what it is to know, to be, to act, and to be an embodied self in time. Western approaches have so far tended to only nibble around the edges of these traditions. . . . What the meditation traditions have to offer science is not just more data to plug into the old ways of looking at brains, but a whole new way of looking.[30]

A radically inclusive empiricism. True science takes all empirical results seriously and does not preemptively exclude certain data because its interpretation may be too complicated, or because there is no currently known theoretical explanation. In the present context, it makes great sense to take seriously observations relevant to mind and consciousness that might promote a deeper analysis of the subject. Many of the phenomena discussed in this book are at the edge of scientific knowledge—puzzling, and lacking an accepted theoretical explanation. Closer investigation may suggest directions toward an expanded science of mind.

Another of William James's lifelong interests was scientific investigation of anomalous psychological phenomena, the existence of which would be strongly relevant to the mind-body relation. These included telepathic communication between individuals, mediumistic communication, and other phenomena suggesting that mind and consciousness, under some conditions, transcend the boundaries of the physical body in inexplicable ways.[31] James believed that to get to the next level of investigating mind in Western science, one would need to take seriously all verified empirical phenomena, no matter how strange and inexplicable they might appear.

In 1892, William James wrote this about the future development of a science of mind:

> A genuine glimpse into what it is would be *the* scientific achievement, before which all past achievements would pale. But at present psychology is in the condition of physics before Galileo and the laws of motion, of chemistry before Lavoisier and the notion that mass is preserved in all reactions. The Galileo and the Lavoisier of psychology will be famous men indeed when they come, as come they some day surely will, or past successes are no index to the future. When they do come, however, the necessities of the case will make them "metaphysical." Meanwhile the best way in which we can facilitate their advent is to understand how great is the darkness in which we grope, and never to forget that the natural-science assumptions with which we started are provisional and revisable things.[32]

James's words capture the notions that developments in the scientific exploration of mind will eventually be truly revolutionary, eclipsing all prior scientific achievements in impact; that the implications will necessarily be metaphysical, meaning that a physicalist metaphysics as currently understood may no longer be sufficient as the explanatory framework; and that this may be accompanied by significant revision of the underlying laws of physics (the "natural-science assumptions"). The next really big scientific revolution may encompass both mind science and physical science, interconnecting the two domains in new and unexpected ways.

From colliding billiard balls and revolving planets of Newtonian classical mechanics, to abstract mathematical descriptions of electromagnetic fields, to bending of space-time geometry by mass and energy, to nonlocalizable wave particles of quantum mechanics, the picture of reality provided by physical science has been over the centuries an evolving scenario. There is little reason to expect that significant revision of our descriptive framework will not continue to take place.

The phenomena described in the following chapters are seemingly not explicable within the currently existing conceptual framework of physical and biological science. However, they can still be studied responsibly using the methods of science. And there is enormous value in doing science in places where the answers are not known and the phenomena are deeply puzzling. Physicist Richard Feynman (1918–1988) put it like this:

> It is only through refined measurements and careful experimentation that we can have a wider vision. And then we see unexpected things: we see things that are far from what we would guess—far from what we could have imagined. . . . If science is to progress, what we need is the ability to experiment, honesty in reporting results—the results must be reported without somebody saying what they would like the results to have been. . . . In fact it is necessary for the very existence of science that minds exist which do not allow that nature must satisfy some preconceived conditions. . . . One of the ways of stopping science would be only to do experiments in the region where you know the law. But experimenters search most diligently, and with the greatest effort, in exactly those places where it seems most likely that we can prove our theories wrong. In other words we are trying to prove ourselves wrong as quickly as possible, because only in that way can we find progress.[33]

May it be the case that our scientific enterprise is up to the task of truly honoring these words. A careful look at the phenomena described in the chapters to follow may take us as scientists—as seekers of knowledge—to some very exciting new places.

2

NEAR-DEATH EXPERIENCES

BRUCE GREYSON

N ear-death experiences (NDEs) are profound events that many people report when they come close to dying. Most people who come close to dying do not report NDEs, but studies in several different countries have confirmed that between 10 and 20 percent of people who have a documented cardiac arrest will describe an NDE.[1] Although the prototypical NDE occurs when someone who is having a heart attack watches the body being resuscitated, there are a great many NDEs reported by people who come close to death psychologically, but whose hearts are not known to have stopped— for example, when they have had a car accident or when they nearly drowned.

In fact, the first large collection of NDEs was composed of accident cases that did not involve cardiac arrest. Swiss geology professor Albert von St. Gallen Heim (1849–1937) had a profound NDE during a fall while he was climbing in the Alps in 1872. As he fell, his body crashed repeatedly against the rocky cliffs. He wrote that he had watched people fall previously, and found it to be a terrifying experience, but that when he himself was falling, it was to his surprise a beautiful experience. He reported being astounded that he was feeling no pain at all.

Heim was so affected by his experience that he started asking other climbers who had survived falls, and he quickly found thirty other cases similar to his. He published this collection of cases in the *Yearbook of the Swiss Alpine Club* in 1892.[2]

The term "near-death experience" (NDE) was coined by French philosopher Victor Egger (1848–1909)—"experiences de morte imminente"—and the phrase was popularized in English by Raymond Moody in 1975. Moody pioneered the contemporary study of NDEs by collecting 150 case reports and abstracting from them common characteristics of the experience. His book on the subject, *Life After Life*,[3] achieved widespread readership. In the several decades since, a small number of researchers have conducted extensive empirical investigations of NDEs.[4]

CHARACTERISTICS OF NDES

The typical features of NDEs include changes in thinking, changes in emotions, seemingly paranormal elements, and features that seem to be otherworldly. The changes in thought processes include distortions in the sense of time, often described as a sense that time had ceased to exist; thinking becoming much faster and clearer than usual; a flood of memories from the person's entire life, sometimes in the presence of a guide who helps the experiencer evaluate what was important and meaningful in his or her life and what was not; and a sense of revelation or sudden understanding.

The emotional changes include an overwhelming sense of peace and well-being, as if whatever is happening is exactly what is supposed to happen, often accompanied by the total elimination of any pain or distress, which is itself remarkable in a near-death situation. The emotional changes also include a feeling of joy, a sense of cosmic unity or becoming one with everything, and a feeling of unconditional love unlike anything the experiencers had felt previously.

The paranormal features include a sense of being outside the physical body. Although some people report just a vague sense of having left the body, others describe accurate perceptions of the body and events around it from an out-of-body visual perspective. The paranormal features also include unusually vivid physical sensations, in which people report seeing colors and hearing sounds that are unlike anything seen or heard in their earthly lives; becoming aware of events going on at some distance from their bodies, as if they had extrasensory perception; and visions of their personal future or sometimes of the future of the planet.

The otherworldly features include entering some unearthly realm or dimension of existence, encountering other entities, such as deceased loved ones or religious figures, and, in some of the more elaborate cases, an entity they experience as a divine or ultimate presence; and coming to a border or point of no return, beyond which they would not be able to return to earthly life.

Although NDEs with all these features have been reported by people whose hearts have not stopped, a study of patients admitted to a cardiac care unit found that NDEs were reported ten times more often by those who survived definitive cardiac arrest than by patients with other serious cardiac incidents.[5] And although all elements of the NDE can be reported by individuals who merely believe themselves to be near death, an encounter with a brilliant light, enhanced cognitive functions, and positive emotions are more common among individuals whose closeness to death has been corroborated by medical records.[6]

Researchers have not identified personal traits or variables that can predict who will have an NDE or what kind of NDE a person may have. People who have had an NDE are indistinguishable from other survivors of close brushes with death in terms of age, gender, race, intelligence, neuroticism, extroversion, anxiety, history of mental illness, or personality traits. We have reports of NDEs from almost every culture across the globe that go back centuries. Although the term "near-death experience" was not widely used until 1975, accounts of these events can be found in the folklore and writings of European, Middle Eastern, African, Indian, East Asian, Pacific, and Native American cultures. The ones that were recorded two thousand years ago and from different cultures are very similar to the ones we hear today from critically ill patients in any hospital in the United States.[7]

There are some cross-cultural differences in how people interpret these experiences. For example, people all over the world will report seeing a warm being of light that makes them feel welcomed and loved. But because that welcoming being is often a figure without discernible features, it may be variously identified as God or Christ, or as a familiar yogi, or as a deceased loved one, depending partly on the experiencer's cultural and personal background and past experiences.[8]

There are other cross-cultural differences as well that suggest prior beliefs have some influence on the kind of experience a person will report following a close brush with death. For example, while encountering other beings and other realms are nearly universal elements, reviewing one's life and traveling through a tunnel are primarily reported from Christian and Buddhist cultures, but are rare among indigenous populations in North America, Australia, and the Pacific Islands.[9] These cultural influences may reflect not so much the experience itself as the individuals' ability to process and describe an experience that is largely ineffable and, if it is to be shared with others, must be put into the images, concepts, and symbols available to the experiencer's culture.[10]

PROLONGED AFTEREFFECTS OF NDES

One of the most important things to me, as a psychiatrist, about these experiences is how they change people. People are often profoundly affected by these experiences for the rest of their lives. I have talked with people in their nineties who had the experience as children and they say that the aftereffects are as strong now as they were many decades earlier. Experiencers typically report increases in spirituality, compassion and concern for others, altruistic behavior, sense of connection to others, sense of meaning and purpose in life, and appreciation for life; and decreases in fear of death, materialism, and competitiveness. The most pervasive change is that people are not afraid of death after the experience. They say they know with certainty that death is not the end, that there is persistent consciousness after the body dies. Although decreased fear of death sometimes increases the risk of suicide, people who have had NDEs paradoxically become more strongly opposed to suicide, primarily because of their increased sense of meaning and purpose in life.[11]

Experiencers tend to see themselves as integral parts of a benevolent and purposeful universe, in which personal gain, particularly at others' expense, is no longer relevant. They typically return with a sense that we are not alone in the world, that we are all part of something greater. They also report being less addicted to worldly things—to money, to power, to prestige; but that does not mean they are uninterested in this world. In fact, they tend to enjoy life more, because they are no longer afraid of dying, and therefore less afraid to take risks and act spontaneously. They tend to become more mindful of things in this world; and they tend to approach this world with a beginner's mind. Every apple they eat is as if it were their first, and they enjoy the sensations of it more than they did before. In sum, they emerge from an NDE with a decreased fear of death, an increased appreciation for life, and a more mindful approach to life.[12] These personal transformations associated with NDEs go beyond what we see in people who have come close to death without an NDE. For example, although many individuals who come close to death express greater appreciation for life than they used to, those who do not have NDEs often become more anxious and depressed, withdraw from social activities, and have posttraumatic stress symptoms.[13]

CAUSAL UNDERSTANDING OF NDES

Various explanations have been proposed to account for NDEs. Some have suggested that NDEs are products of the imagination, constructed from our personal and cultural expectations and hopes, to protect us from facing the threat of death. That is, NDEs are nothing more than wishful thinking, with the dying persons psychologically conjuring up the images of what they would want to happen at death. However, research does not support this idea. Studies in several different countries have shown no association between religion or religiosity and NDEs, and people often report experiences that conflict with their specific religious and personal expectations of death. Furthermore, people who have never heard or read of NDEs describe the same kinds of experiences as do people who are quite familiar with them; and the knowledge that people had about NDEs prior to their experiences does not seem to influence the details of their NDEs.[14] NDEs described prior to 1975, when the phenomenon was first publicized, are essentially the same as those experiences described by people today.[15] Furthermore, children too young to have received substantial cultural and religious conditioning about death report the same kinds of NDEs as do adults.[16]

Several physiological explanations have also been proposed for NDEs. Because many NDEs happen in hospitals, there are physiological data bearing on these proposed explanations. For example, some have speculated that lack of oxygen to the brain plays a major role in NDEs because, no matter how one comes close to death, reduced oxygen flow to the brain is usually a final common event. But studies have shown that people who report NDEs actually have more oxygen getting to their brain than do those people who do not report NDEs after a close brush with death.[17]

Others have speculated that NDEs are caused by drugs given to people who are dying. But various researchers have looked at drugs given to people who are dying, and found, in fact, that the more drugs people are given, the fewer NDEs they report.[18] Similarly, people who have brain malfunctions, such as those due to high fevers, report fewer NDEs than people who are more stable metabolically when they come close to death. Furthermore, hallucinations caused by brain malfunctions generally produce confusion, irritability, fear, and idiosyncratic, bizarre visions. These hallucinations are not at all like the exceptionally clear thinking, peacefulness, calmness, and consistent visions generally seen in NDEs.

Certainly there may be physiological and chemical changes in the brain that are related to some aspects of the NDE. Some have speculated that these experiences are caused by endorphins or other chemicals released in the brain under stress as someone is dying. Perhaps endogenous brain chemicals do play a role; only additional research can address this. However, to date, there is no known neurochemical or drug substance that will reliably produce the various characteristics of NDEs.[19]

Surprisingly little is known about the physiology of dying, including very little about the neurophysiology of the dying brain.

There have been very few measurements of brain electrical activity at the time of death. In those few instances, both human measurements recorded via electrodes placed on the forehead[20] and recordings obtained using implanted electrodes in rats subjected to cardiac arrest have indicated some brief persistence of brain electrical activity at the time of death.[21]

The human studies of postmortem electrical activity did not use a standard electroencephalograph (EEG) but rather a proprietary index of electrical activity detected on the forehead, which has been used to monitor depth of anesthesia but has no known utility in measuring conscious information processing,[22] and is known to misinterpret a variety of physiological and environmental electrical signals as brain activity.[23] This statistical measure of electrical activity based on recordings from the patient's forehead is particularly vulnerable to false readings coming from electrical activity of the underlying forehead muscle, which even at rest can produce signals that resemble brain waves.[24] Standard EEG recording in humans, by contrast, invariably shows brain activity decreasing and disappearing entirely within ten to twenty seconds of cardiac arrest, without any surge in activity.[25]

In the rat study, a specific increase in synchronous higher frequency activity (gamma: ~30 hertz and greater) was detected. As these high-frequency synchronous oscillations have been proposed as neural correlates of conscious awareness,[26] it was contended that such neural activity may have something to do with the vivid experiences reported in NDEs. However, drawing conclusions about human mental activity from rat physiology is problematic; the postmortem electrical activity observed in these rats was only a small fraction of the total neuro-electric power present before death of the rats, which does not suggest complex cognitive processing; and the increased electrical activity was not observed in rats who had been anesthetized, whereas NDEs are common in anesthetized humans.[27] This surprising electrical activity around the time of death therefore appears not to be correlated with the complex mental activity seen in NDEs. Nevertheless, the possibility that NDEs may be associated with neural activity near death is an interesting hypothesis that merits further investigation.

Some have compared NDEs to temporal lobe seizures since there are some superficial similarities; and some researchers have claimed that they can induce a sensation of being out of the body by stimulating the temporal lobe of the brain.[28] But these sensations induced by temporal lobe stimulation differ in several important ways from spontaneous out-of-body experiences (OBEs), such as those that occur in NDEs.[29] Furthermore, showing an association between the temporal lobe and a feeling of leaving the body would not imply that all OBEs are caused by temporal lobe malfunctions. Stimulating the temporal lobe can also produce hallucinations of hearing music, but that does not mean that every time you listen to music you are just hallucinating. When you listen to music produced outside your brain in the real world, your perception of it is processed in the temporal lobe. Likewise, if you were to leave your body in reality, your brain might process those perceptions in the temporal lobe. Rather than creating the experience, your brain might be channeling it, much like a television set or a radio. Understanding how a television set works does not say much about where the images and sound originated; and understanding how the brain processes experiences does not necessarily reveal anything about the source of the experience.

In fact, the vast majority of experiences triggered by stimulating the temporal lobes bear no resemblance at all to NDEs, and the vast majority of people with temporal lobe seizures do not report OBEs.[30] In our research in the epilepsy clinic at the University of Virginia, we found that OBE-like episodes were extremely uncommon during seizures and, when they did occur, were not preferentially associated with the temporal lobe; and that NDE-like episodes were nonexistent among the patients surveyed.[31]

EXPERIENCES ASSOCIATED WITH THE MINIMALLY FUNCTIONING BRAIN

The main problem with all these physiological and psychological explanations of NDEs is that they assume a brain that is functioning well enough to visualize or fantasize and form coherent memories. But when the brain is working well below capacity, as it is in most NDEs, there should not be any clear thinking or memory formation. Almost all of the people who have had NDEs report that, during the experience, their thinking was clearer, faster, and more logical than ever before, and those reports are difficult to reconcile with a brain that is not functioning well.

Cardiologist Michael Sabom[32] described in detail the NDE of a woman with an aneurysm in her brain that was so large and so hard to reach that it could not be removed by standard neurosurgical techniques. Instead, she underwent a procedure called hypothermic cardiocirculatory arrest, a drastic operation in which she was connected to a cardiopulmonary bypass machine, her body temperature lowered to 60°F, and her heartbeat and breathing stopped. At this point, the surgical table was tilted up and the blood drained from her body so that the aneurysm could be safely removed. During this procedure, she met the accepted criteria for brain death: her EEG (electroencephalograph: a measure of brain electrical activity, resulting from neural signaling) was totally flat, indicating no electrical activity in her cerebral cortex; and she had no auditory evoked potentials, indicating lack of neural activity in her brainstem.

When she awoke from the anesthesia, she reported a profound NDE in which she saw accurately, from a position above her body, surprising details of events that occurred during her operation. Details of her reported perceptions during the operation were verified by participating surgical personnel. Though the veridical perceptions of her physical environment occurred while she was deeply anesthetized, they did not refer to events during the period of the documented flat EEG, when the blood was drained from her brain and she would have met the clinical criteria for brain death. Still, there are many inexplicable things about reports such as these stemming from states of deep anesthesia, and additional careful studies of experiences that occur in conjunction with monitored surgical procedures is a valuable direction for future research.[33]

Recently, a survey was conducted of patients who had undergone a hypothermic cardiocirculatory arrest procedure over a five-year period.[34] Nine percent of the surveyed patients reported conscious mental activity during the time when their brains were not functioning, and one case was described in which the patient felt peace and joy, saw a bright light, left her body, and described details of the operation that were later verified by the surgical staff. The question raised by these cases and many others is how we can reconcile a brain that is not functioning with a mind functioning better than ever.

There are two contrasting and long-standing analogies that have been proposed as models of the mind functioning when the brain starts to fail. Two nineteenth-century scholars laid them out clearly: British neurologist John Hughlings Jackson and

German philosopher Carl du Prel. Jackson's "perceptual release model" is illustrated by the analogy of sitting in a room with a fire blazing in the fireplace, and a picture window looking out into a garden. In the daylight, one sees the outside garden clearly through the picture window. As the sun goes down and there is less light coming in from outside, one no longer sees the garden outside through the picture window, but rather the reflection of the fire in the fireplace. By this analogy, we lose contact with the outside world as the brain shuts down, and what we see instead are reflections of what is going on inside the brain—that is, hallucinations.[35]

The contrasting model, described by du Prel, used the analogy of looking up at the sky. During the day, one sees the bright sun and little else. As the sun goes down, it gets darker and darker, and one sees the moon, stars, and planets. By this analogy, as the brain shuts down, it ceases to block out external phenomena that were always there but overshadowed by brain processes.[36] We have these two different models of why unusual sensory experiences become evident when the brain stops functioning. Is the brain creating hallucinations, or are we gaining access to other domains to which we are usually blinded by the functioning brain?

ENHANCED MENTAL CLARITY

There are some features of the NDE that do not seem compatible with a physiological explanation in terms of brain function. One is simply the exceptional mental clarity during NDEs—people report that their thinking was clearer than ever, despite their brain function being severely compromised. Among the several hundred people who have had NDEs that we have studied, 80 percent describe their thinking as clearer than usual or at least as clear as usual. In addition, 74 percent described their thinking as faster than usual or at least as fast as usual; 65 percent described their thinking as more logical than usual or at least as logical as usual; and 55 percent described their control over their thoughts as more control than usual or at least as much control. Furthermore, an analysis of their medical records showed that mental functioning was significantly better in those experiencers who had come closest to death. This enhanced mental functioning while the brain is impaired suggests that there is more going on with the mind than its being a simple consequence of brain physiology.[37]

One experiencer whom I interviewed had attempted suicide by overdosing.[38] He took what he thought was a lethal number of pills and lay down on his bed to die. Instead of dying, he got sicker and sicker, more and more confused and, eventually, had trouble breathing. This was clearly more pain than he had bargained for, and he decided he should call for help. He got up out of his bed and tried to make his way to the telephone. He could not do that, partly because he was staggering too much, and partly because he was hallucinating many little people milling around his room, preventing him from reaching the telephone. As he was looking around confused at all these little people, wondering who they were and what they were doing in his apartment, and finding it harder and harder to breathe, he left his body. He rose up about ten feet behind the body and he looked down at the scene. From his perspective, he was thinking clearly; he looked down and saw his body, still standing and looking around in confusion. From his vantage point, he could not see the hallucinations his body was looking at. He remembered being in the body, and he remembered that his body was seeing imaginary little people, but he could not see them where he was. In this case, his brain was hallucinating, but his mind was thinking clearly.

Another example of enhanced mental functioning during NDEs is the rapid revival of memories that sometimes includes events from the person's entire life. Among a large sample of people who have had NDEs that we have studied, 13 percent reported a life review during the NDE. Moreover, in contrast to the isolated and often just single brief memories that can be induced by stimulation of the temporal lobes, 59 percent of people who have had NDEs who reported a revival of memories during the NDE said they had experienced an almost instantaneous panoramic review of every event in their entire life.[39]

OUT-OF-BODY EXPERIENCES (OBES)

A second feature of NDEs that physiological explanations have difficulty accommodating is accurate perception from an out-of-body perspective. There are many documented examples of people who claim to have left their bodies and seen things from a visual vantage point that they could not have seen if they were still in the body, even if the body were awake and functioning. Among the several hundred people who have had NDEs that we have studied, 81 percent reported feeling separated from their physical bodies during the NDE and 48 percent reported accurate out-of-body perceptions.

One example was a fifty-year-old truck driver who had quadruple bypass surgery, in the midst of which he found himself above his body, looking down. He described the cardiac surgeon "flapping his arms as if trying to fly." This was quite unusual behavior for a surgeon in the midst of an operation, and not the type of behavior portrayed on television shows about doctors.

When he asked the surgeon about it at his subsequent follow-up visit, the surgeon got defensive and asked, "Who told you about that?" The patient said, "Nobody told me. I was watching from above. I died and left my body, and I was watching from above." The surgeon then said only, "Well, you're here, you're alive, so I must have done something right!" That was the end of the conversation. A few years later, I talked with the surgeon about this incident. He acknowledged that the patient's observation was accurate. He had scrubbed and gowned and gloved, and entered the operating room "sterile." He let his residents start the surgery, and they started cutting the patient's chest open. As the surgeon was supervising their technique, he did not want to risk touching anything with his sterile gloves, so he put his hands where he knew they would not touch anything: flat against his chest. Then he supervised them, pointing with his elbows as he directed them where to cut. His directing with his elbows did make it look like he was trying to fly.[40] The question is how the patient who was unconscious from anesthesia could see this from above the operating table, unless he was in fact outside his body.

Another example, not from our case collection but reported by Kimberly Clark Sharp, was a migrant laborer who was brought into a hospital unconscious after a heart attack. She was resuscitated in the emergency room and admitted to the cardiac care unit. When she eventually regained consciousness, she was very excited and told staff she had left her body and risen above the hospital. She reported that she had seen, on the ledge outside of a window on an upper floor on the far side of the hospital from where her room was, a blue tennis shoe. She could not figure how it got there. She described it in detail, including scuffmarks on the toe and the fact that the shoelace was tucked under the heel. Sharp, the hospital social worker, took her seriously enough to look for the shoe. She went from room to room, looking out the windows, and finally found the shoe on a ledge on the far side of the hospital, exactly as the patient had described it.[41]

Some people who have had NDEs furthermore report that while they were out of their bodies they became aware of events occurring at a distance, or that in some other way would have been beyond the reach of their ordinary senses even if they had been awake and functioning normally. We have published a review of fifteen such accounts that included accurate perceptions of unexpected or unlikely details.[42] Additionally, Kenneth Ring and Sharon Cooper reported a study of thirty-one blind people who have had NDEs, many of them blind from birth, who were able to describe the scene around them while they were out of their bodies during their NDE. Some of their accounts even included accurate descriptions of the colors of specific items.[43]

Michael Sabom, the cardiologist who reported the hypothermic cardiocirculatory arrest case described above, asked a series of resuscitated patients who reported NDEs to describe the scene they had observed in their experiences. He also asked a matched group of "seasoned cardiac patients" who did *not* have NDEs during their cardiac crises to report what they *imagined* the scene might have looked like if they *had* been watching from out of their bodies. None of the patients *without* NDEs could describe the scene accurately; and 80 percent made major errors in their imagined scenarios. In contrast, none of the people who have had NDEs made *any* errors in their descriptions, and 19 percent reported specific idiosyncratic and unexpected events that happened during their resuscitations.[44] A similar study was carried out in Wales by Penny Sartori, an intensive care nurse, with similar results: the people who have had NDEs described unusual and surprising details of their resuscitations accurately, whereas every patient who did not have a NDE made major errors in their descriptions.[45]

Janice Holden recently reviewed ninety-three reports of potentially verifiable out-of-body perceptions during NDEs.[46] These did not rely solely on the experiencers' testimony; in fact, she found that 86 percent had been corroborated by an independent witness. Of these out-of-body perceptions, 92 percent were completely accurate in every detail, 6 percent contained some minor error, and only one was completely erroneous. Even among those cases that were corroborated to the investigator by an independent witness, 88 percent were completely accurate. For example, Dutch cardiologist Pim van Lommel and his colleagues reported a case in which a cardiac arrest victim was brought into the hospital emergency room unconscious and remained in a coma and on artificial life support in the intensive care unit. When he finally regained consciousness more than a week later, he immediately recognized one of the nurses as the person who had removed his dentures during the resuscitation procedure in the emergency room, and he described correctly and in detail the entire procedure, including the cart in which the nurse had put the dentures, which none of the hospital staff had been able to find in the ensuing week.[47]

Some writers skeptical of the anomalous nature of NDEs have suggested that the accurate perceptions reported by experiencers actually occurred while the brain was going offline or coming back online, rather than during the period of minimal or no brain activity.[48] As noted above, Sabom's patient who had an NDE during hypothermic cardiocirculatory arrest reported accurate perceptions of her physical environment while she was anesthetized, but not during the period of the documented flat EEG. However, some of the ninety-three NDEs reviewed by Holden, including the one described above reported by van

Lommel and colleagues, are clearly anchored in time to a period of minimal or absent brain activity, rendering them inexplicable in terms of conventional physicalist neurobiology.

Some skeptics dismiss these cases as "anecdotes," implying that they are worthless and can be ignored, as if they did not happen. Retrospective accounts of these uncontrolled percep-tions could in some cases be attributed to unconscious awareness of sensory cues, subliminal perception, memory distortion, and erroneous estimates of the probability of the reported events. It is true that anecdotes, though far from worthless, are not in fact as evidential as controlled experiments. However, controlled experiments are not often feasible because most NDEs happen in places and under circumstances not under our control.

Nevertheless, there have been a few attempts to provide con-trolled evidence by placing visual displays hidden from direct sight, near the ceiling, facing upward, in hospital settings. Five different hospital-based research teams carried out small stud-ies of this type that altogether included a total of only twelve NDEs with out-of-body perceptions, none of which included descriptions of the visual displays that had been designated (unknown to the patients) as the "targets."[49] An international consortium spearheaded by Sam Parnia recently reported results of a four-year multicenter study of awareness during car-diac resuscitation, including attempts to assess whether patients had veridical out-of-body perceptions.[50] In their sample of more than two thousand cardiac arrests that occurred in the hospital, only one hundred forty patients survived, were medi-cally fit enough to be interviewed, and agreed to be interviewed. Among those one hundred forty patients, nine reported having NDEs, of whom two reported visual perceptions of the envi-ronment during the period of their cardiac arrest. One of those patients was too ill to provide an account of his perceptions, but the other described detailed, accurate perceptions of events that were anchored in time to the period of his cardiac arrest. Unfortunately, neither of these cardiac arrests happened to occur in a hospital room in which visual targets had been placed. The study thus provided further evidence of veridical out-of-body perceptions during a period of minimal brain function, but not of perception of the specific visual display designated (again, unknown to the patient) as the target. Obviously, such planned prospective studies of OBE perception are very difficult and require exceptional time and resources, but research in this area continues to hold promise.

ENCOUNTERS WITH DECEASED INDIVIDUALS

Another feature of NDEs that is hard for physiological models to explain is the apparent encounter with deceased people. Many people who have had NDEs report that, during the time they seemed to be dying, they met deceased relatives and friends. Among several hundred people who have had NDEs that we have studied, 42 percent reported meeting recognizable deceased acquaintances in their NDEs. These encounters were more likely to be reported the closer the experiencer had come to death. In some of these cases, the purported deceased individuals provided information that the person who had an NDE could not have known otherwise. The most common skeptical explanation for apparent encounters with deceased loved ones is that they are due to wishful thinking. That may, in fact, be a plausible explanation for some of these cases, but in 33 percent of the cases we have studied, the deceased person seen was not someone the experiencer wanted or expected to see, but was instead someone else.[51] Wishful thinking is also an unlikely explanation in the NDEs of children, who we would expect to wish to see their living parents or other protectors in times of stress. In fact, however, children virtually never see their living parents in NDEs. In some cases, children who have had NDEs described meeting people they did not recognize, but described them in sufficient detail to allow their parents to recognize them as deceased relatives whom the child had never met.[52] There is furthermore another kind of NDE in which expectation and wishful thinking cannot possibly account for seeing deceased loved ones, and that is when the person who had an NDE could not have known that the deceased person encountered had died. A number of such cases have been described, some of them quite well documented.[53]

The earliest such case of which I am aware was reported by Pliny the Elder in the year 77 CE. He wrote that a Roman nobleman named Corfidius had stopped breathing and was assumed to be dead. His younger brother took him to a mortician to be prepared for burial. Later, however, Corfidius revived on the embalming table, much to the undertaker's horror. He told the undertaker that he had just been to his younger brother's house, and discovered that his younger brother had just died. He claimed that his deceased younger brother told him that he had already paid for a funeral and that that payment should now be used for his own burial instead. Corfidius also claimed that his deceased younger brother told him about some buried money in his yard that no one knew about. As he was relating this account to the undertaker, his younger brother's servant ran into the mortuary with news that his master had just died.[54]

A contemporary case of this kind involved a nine-year-old boy in Pennsylvania named Eddie, who had severe meningitis. Eddie hovered near death for thirty-six hours before his high fever finally broke. His parents did not think he was going to live through the night, and they held an all-night vigil by his bedside in the hospital intensive care unit. Around three o'clock in the morning, Eddie woke up, very excited, wanting to talk about the experience he had just had in heaven. He said that he had seen his deceased grandfather, aunt, and uncle. To his family's consternation, he claimed that he also met his nineteen-year-old sister Teresa. Teresa was in college in Vermont, and as far as the family knew, was alive and well. But Eddie insisted that she was there in heaven and had told him he had to go back. His father got increasingly upset at Eddie's apparent delusional ramblings about Teresa, and finally persuaded the doctor to sedate Eddie to stop him from ranting. The parents then went home, convinced that Eddie was safe now, and they found a message on their answering machine from Teresa's college, saying that she had been in a terrible automobile accident. They called the college and found she had died around midnight.[55]

In addition to these NDEs involving encounters with deceased loved ones who were not known to have died, and which therefore cannot be dismissed as due to expectation, there are also cases in which people who have had NDEs encountered a deceased person they did not know. For example, one young girl undergoing open heart surgery met a boy who claimed to be her brother, although she thought she was an only child. When she recovered from the surgery, she asked her father about this. He acknowledged that she had, in fact, had a brother who had died before she was born, and whom they had never mentioned to her.[56]

Pim van Lommel wrote about a man in the Netherlands who had an NDE during heart surgery in which he saw a man he did not recognize but who looked on him very lovingly. When he recovered, he asked his parents about this. He described this man to the parents, and they did not acknowledge who that man could possibly be. Years later, his mother, on her deathbed, confessed to him that, in fact, the man who raised him was not his father, that she had been in a relationship with a man who was Jewish and had become pregnant by him, but that man was then captured by the Nazis and never seen again. She then married the person who raised this boy as his own son. She showed him a picture of the man who was his biological father, and he recognized it as the man from his NDE.[57]

CONCLUSION

One prolonged aftereffect of NDEs that has been of particular interest to me is the influence of such experiences on people's thinking about life and death. I recently interviewed a patient at the University of Virginia Hospital who had attempted suicide. He had made his suicide attempt very reluctantly, because he was afraid he would be condemned to hell if he killed himself. However, his pain became so intolerable that he did it anyway, and as a result of his close brush with death, he had an NDE. He returned thinking that death was a blissful experience, something to look forward to rather than something to fear. One might expect that that change in attitude might have made him more suicidal, but in fact the effect was just the opposite. In addition to his altered attitude toward death, he also returned with the sense that there is a vital meaning and purpose to his life. He did not know what that purpose was, but he now had the sense that the pain and suffering he had gone through happened for a reason, and that his responsibility was not to try to escape that, but rather to deal with it. Like most people who have had NDEs, he emerged from his NDE with a profoundly spiritual attitude toward his life. He now believed that he was more than just a collection of molecules; but that he was a spirit with meaning and purpose to his earthly circumstances, with a profound connection to everything else in the universe.

To summarize, although NDEs have been reported for centuries, they have become objects of scientific study only recently, as modern technology has allowed us to bring people back from the brink of death and indeed to blur the boundary between life and death.[58] NDEs include phenomena such as mental clarity when the brain is severely impaired, accurate out-of-body perceptions, and visions of deceased people sometimes communicating information that no one else knew. These features challenge the common assumption that consciousness is solely the product of brain processes, or that the mind is merely the subjective concomitant of neurological events, and instead lend support to the alternative view that brain activity normally serves as a kind of filter, selecting the mental content that is allowed to emerge into waking consciousness.

Furthermore, NDEs result in profound changes in attitudes, beliefs, and values that suggest a spiritual as well as a physiological component to the experience. Most people who have had NDEs report dramatic and long-lasting increases in their sense of spirituality, compassion and concern for others, altruistic behavior, appreciation of life, sense of connection to other people and indeed to other life forms as well, belief in life after death, and sense of meaning and purpose. At the same time, they report a decreased fear of death, as well as decreased interest in material things, personal status, and competition. Both the empirical support NDEs provide for a component to humankind that transcends our physical bodies, and the changes in values, beliefs, and attitudes they promote in experiencers, are consistent with the core messages of most enduring spiritual traditions.

3

REPORTS OF PAST-LIFE
MEMORIES

JIM B. TUCKER

For more than fifty years, researchers at the University of Virginia have studied cases of children from various parts of the world who report memories of previous lives. These cases typically involve children who spontaneously talk about a previous life at a very young age. No hypnosis is involved, and the children do not talk about being royalty or famous people from the past. Instead, they describe ordinary lives, usually saying they lived fairly close by, sometimes in the same village and almost always in the same country. The lives they report are recent ones, with the median time between the death of the previous person and the birth of the child being only sixteen months. Some of the children talk about being deceased family members such as a grandparent. Others describe being strangers from a different location. If the child gives the name of the place, then people have sometimes gone there and found that, in fact, someone did live and die whose life matches the details the child has given.[1]

HISTORY OF THE RESEARCH

Ian Stevenson (1918–2007) largely created a new area of scientific research when he began studying cases of children who report memories of previous lives. He came to the University of Virginia to be the chair of the Department of Psychiatry in 1957. At the time, he had already published extensively in the field of biological psychiatry, and he also had an interest in parapsychology. When the American Society for Psychical Research[2] announced a contest for the best essay on paranormal mental phenomena and their relationship to the question of survival after death, he submitted what turned out to be the winning entry. In it, he reviewed forty-four published reports of individuals, primarily young children, who had reported memories of previous lives.[3]

After Stevenson's paper was published, Eileen Garrett, a well-known medium who was president of the Parapsychology Foundation, contacted him.[4] All of the reports in Stevenson's paper had involved past cases, but Garrett had learned of a child in India who was making the same kind of statements at that time. She offered Stevenson a small grant to go investigate the case. He accepted and traveled to India in 1961. By the time of the trip, he had learned of five cases. Once he got to India, he found twenty-five cases. Similarly, after hearing of one or two cases in Ceylon (Sri Lanka), he spent a week there and saw seven cases. He realized that children's claims of past-life memories were much more common than anyone in the West had known. Thus began an interest that would consume much of Stevenson's professional life for the next forty years.

Along with Eileen Garrett, another person who read Stevenson's article with great interest was Chester Carlson (1906–1968), the inventor of xerography, the photocopying process that became the basis for the Xerox Corporation. Carlson offered to fund research into cases like the ones Stevenson described. Stevenson initially turned him down because of his obligations as chair of his department, but as he became more intrigued by the cases, he began devoting more time to them with Carlson's help.

Stevenson published his first book on the topic, *Twenty Cases Suggestive of Reincarnation*, in 1966.[5] The title typifies Stevenson's cautious approach. Rather than accepting the cases at face value, he attempted to document them as carefully as possible, at times in exhaustive detail. Stevenson considered himself a man of science and always strove to use a scientific attitude, being careful and methodical in trying to sort out exactly what had happened in the various cases. He never assumed they meant reincarnation had occurred, and that is the approach we still use today.

With Carlson's funding, Stevenson stepped down as chair of the department in1967 to focus full-time on research. He created the research division now known as the Division of Perceptual Studies (DOPS) in which to carry on the work. The following year, Carlson died unexpectedly. Though Stevenson feared this would mean the end of his research on this topic, it turned out that Carlson had bequeathed a considerable amount of money to the University of Virginia to support it. Thus, the work continued.

In 1975, Stevenson began a four-volume series of books called *Cases of the Reincarnation Type*. The different volumes included carefully documented cases from India,[6] Sri Lanka,[7] Lebanon and Turkey,[8] and Thailand and Burma.[9] When the first volume was published, the book review editor of *JAMA: The Journal of the American Medical Association*, wrote:[10] "In regard to reincarnation he has painstakingly and unemotionally collected a detailed series of cases from India, cases in which the evidence is difficult to explain on any other grounds."

Stevenson also began to get other researchers interested in the work. These included Satwant Pasricha, a clinical psychologist in India who first assisted Stevenson when she was a student; Erlendur Haraldsson, a psychologist at the University of Iceland; Antonia Mills, an anthropologist who was at the University of

Virginia before going to the University of Northern British Columbia; and Jürgen Keil, a psychologist at the University of Tasmania. They investigated and published cases independently of Stevenson. In 1994, Mills, Haraldsson, and Keil published a paper with a combined collection of 123 cases. They stated:[11] "The investigations of three independent researchers into reported cases of reincarnation in five cultures in which such cases are reported suggest that some children identify themselves with a person about whom they have no normal way of knowing. In these cases, the children apparently exhibit knowledge and behavior appropriate to that person."

Although this work did not gain widespread acceptance, it did garner respect in some mainstream quarters. Carl Sagan (1934–1996), the astronomer and a founding member of what is now called the Committee for Skeptical Inquiry,[12] wrote in 1996:[13] "At the time of writing there are three claims in the [parapsychology] field which, in my opinion, deserve serious study," one being "that young children sometimes report details of a previous life, which upon checking turn out to be accurate and which they could not have known about in any other way than reincarnation."

Stevenson continued to write papers and books about the cases, almost until the very end of his life in 2007. Since his death, others have continued the study of this phenomenon, and we now have over twenty-five hundred cases in our files. While cases are easiest to find in places where most people believe in reincarnation, cases have been found wherever anyone has looked for them. They have been found on all the continents except Antarctica, and they certainly occur in the West as well as in Asia. At the University of Virginia, we have recently been focusing on American cases. They are harder to find, but it is unclear if this is because they are actually less common or because families do not talk about them openly here the way people do in many other cultures. We now have numerous American examples, often occurring in families in which there was no belief in past lives before a child began reporting memories of one.[14]

THE CASE OF KUMKUM VERMA

An example of one of Stevenson's cases is a girl named Kumkum Verma who was from a village in India.[15] When she was three years old, she began saying that in her previous life she had been a woman in Darbhanga, a city of two hundred thousand people that was twenty-five miles away. Along with the name of Darbhanga, she gave the name of the city district where she had lived, one primarily made up of artisans and craftsmen. Her parents did not know anyone there. She talked about this life a great deal, and her aunt wrote down some of what she said before anyone attempted to find whether there had been a previous person whose life matched the given details. Stevenson obtained a partial record of the aunt's notes and had it translated. It listed eighteen statements Kumkum had made, including the name of her son from the previous life and the fact that he worked with a hammer, the name of her grandson, and the town where her father had lived. It included specific personal items Kumkum had described: an iron safe in her home, a sword hanging near the cot where she slept, and a pet snake to which she had fed milk.

Her father eventually talked about this with a friend who had an employee from Darbhanga. That man went to the city and found that, in fact, a woman whose life matched all the statements listed above had died a few years before Kumkum was born. The families were complete strangers to each other a remained so even after the deceased woman was identified. Kumkum's father visited the previous family once, but he never allowed Kumkum to go. He was an affluent landowner and was apparently unhappy that his daughter was reporting the life of a blacksmith's wife. Thus, there is certainly no reason to think Kumkum's family encouraged her to say the things she did.

BIRTHMARKS AND BIRTH DEFECTS

Over the years, Stevenson studied cases of children born with birthmarks or defects that corresponded to wounds (usually the fatal wounds) suffered by the deceased individuals whose lives the children appeared to remember. He went to great lengths to verify that the birthmarks or birth defects did in fact match the previous wounds. He always tried to obtain autopsy reports if they were available, along with medical records or police reports. If no written records were available, he would elicit eyewitness testimony about the wounds. For many years, he put off publishing reports as his collection continued to grow. Finally, in 1997, Stevenson produced *Reincarnation and Biology: A Contribution to the Etiology of Birthmarks and Birth Defects*, a 2,200-page, two-volume collection of over two hundred such cases.[16] He also wrote a shorter synopsis of that work, entitled *Where Reincarnation and Biology Intersect*.[17] In these books, he included numerous pictures of birthmarks and defects that were often highly unusual.

The cases included a girl born with markedly malformed mbered the life of a man whose fingers were y with only stubs for fingers on his right hand the life of a boy in another village who lost ight hand in a fodder-chopping machine; and bered the life of a man who underwent skull surgery. She had what Stevenson called the most extraordinary birthmark he had ever seen: a three-centimeter wide area of pale, scar-like tissue that extended around her entire head. He reported that in eighteen cases in which the previous person was shot, the child was born with double birthmarks, ones corresponding to both the entrance and the exit wound on the previous person's body.

The birthmarks and birth defects appear to represent tangible evidence of carryover from a deceased individual. They indicate that this carryover can involve more than just memories and emotions; it can seemingly produce visible physical manifestations on a developing fetus as well, similar to wounds the previous person experienced.

One particular type of case involved what Stevenson termed "experimental birthmarks." These cases, which are found in several areas of Asia, occur when people make a mark on the body of a person who has just died, usually using soot or paste. They typically make a wish or say a prayer that the deceased carry the mark with them to their next life so they can be identified. This is generally done with the expectation that the individual will be born into the same family. Stevenson collected twenty of these cases[18] and Keil and I found eighteen more.[19] Some of the birthmarks were quite impressive. In one case that Keil and I studied, a daughter-in-law marked the body of a woman with white paste, making a mark down the back of her neck with her finger. A boy, the deceased woman's grandson, was then born with a pale area that looked as if someone had made a mark down the back of his neck. The Dalai Lama wrote in his autobiography about a similar case in his family.[20] When his two-year-old brother died, a mark was made on the boy's body with a smear of butter. Another brother was subsequently born with a pale birthmark in the same place where the first body had been marked.

THE CASE OF CHANAI
CHOOMALAIWONG

Chanai Choomalaiwong, a little boy in Thailand, was born with two birthmarks, one on the back of his head and one on the left side of his forehead.[21] When he was three years old, he began talking about a previous life. He said he had been a schoolteacher named Bua Kai and that he had been shot and killed while on his way to school one day. He gave the names of his parents, his wife, and two of his children from that life, and he repeatedly begged his grandmother to take him to his previous parents' home.

Eventually, when he was still three years old, Chanai and his grandmother took a bus fifteen miles to a town near the one where Chanai had said he lived before. After they got off the bus, Chanai led the way to a house where he said his parents lived. The house belonged to an elderly couple whose son, Bua Kai Lawnak, had been murdered five years before Chanai was born, shot as he bicycled to the school where he worked. Chanai identified Bua Kai's parents, who were there along with other family members, as his own. They were impressed and invited him to return. When he did, they tested him by asking him to pick out Bua Kai's belongings from others, which he was able to do. He also recognized one of Bua Kai's daughters and asked for the other one by name. Bua Kai's family accepted that Chanai was Bua Kai reborn, and he visited them a number of times. He insisted that Bua Kai's daughters call him "Father." If they did not, he refused to talk to them.

Stevenson talked with a number of family members about Bua Kai's injuries, and they said he had two wounds on his head after he was shot. His wife remembered that the doctor who examined Bua Kai's body stated that the entrance wound was the one on the back of his head because it was much smaller than the wound on his forehead that would have been the exit wound. These matched Chanai's birthmarks, a small, circular one on the back of his head and a larger, more irregular one toward the front.

PAST-LIFE STATEMENTS

The children tend to be quite young when they begin describing another life, with the average age being thirty-five months.[22] Some talk in a detached, matter-of-fact way, but many show great intensity describing their apparent memories. They may cry every day about missing their previous life and ask repeatedly to be taken to the previous family. Many of these children may talk about these things with great emotion one minute and then run off to play the next. Some appear able to access the memories at all times; others have to be in the right frame of mind, usually during relaxed times. The American parents often say their children typically talk about a past life during long car rides or after baths.

By the time the children are six or seven, most stop saying such things and seem to lose their purported past-life memories. This is around the age when children everywhere typically lose their early childhood memories. In our cases, even though the children usually stop talking about a past life, some later report as adults that they have still maintained at least some of their memories of one.[23]

When the children talk about a past life, they usually focus on the end of the life. Three-quarters of them talk about the mode of death, which is the one part of the life that is often out of the ordinary. The previous person died by unnatural means 70 percent of the time, more than would be expected by chance, and that appears to be an important factor in this phenomenon. The children also talk about people from the end of the previous life. For instance, children describing the life of someone who died as an adult are more likely to talk about that person's children or spouse than about the parents. Though the children recall some earlier items, they generally talk about events at the end of the previous life, acting as if their memory has simply picked up where it left off when the adult died.[24]

THE CASE OF SUJITH JAYARATNE

Sujith Jayaratne was a boy living in a suburb of Colombo, the capital of Sri Lanka, who began showing an intense fear of trucks—and even the word *lorry*, the British word for truck that is part of the Sinhalese language—when he was eight months old.[25] When he became old enough to talk, he described a life in a village named Gorakana that was seven miles away. A monk made notes of conversations he had with Sujith when he was two and a half years old, and Stevenson was able to obtain a translated copy of them. These document that Sujith said he was from Gorakana and lived in the Gorakawatte section of it, that his father was named Jamis and was missing an eye (Sujith pointed to the right eye to indicate that it was the defective one), that he had attended the *kabal iskole* (which means "dilapidated school") and had a teacher there named Francis, and that he gave money to a woman named Kusuma, who prepared string hoppers, a type of food, for him. He implied that he gave money to the Kale Pansala, or Forest Temple. He said two monks were there, and one of them was named Amitha. He also said his house was whitewashed, that its lavatory was beside a fence, and that he bathed in cool water.

The monk then went to Gorakana and found that all of Sujith's statements were true for the life of a man named Sammy Fernando, who died at age fifty after being hit by a truck six months before Sujith was born. When Stevenson investigated the case a year after the monk's trip to Gorakana, he learned that two people in Sujith's neighborhood had connections to Sammy Fernando. Sujith's family knew one of them, a former drinking buddy of Fernando's, slightly, and the other, Fernando's younger sister, not at all. The family had no idea who Sujith was talking about until the monk went to Gorakana—in fact, neither Sujith's mother nor the monk had heard of Gorakana before the case developed.

Along with his statements about Fernando's life, Sujith showed behaviors that appeared connected to it, including an unfortunate interest in arrack, a liquor that Fernando transported and consumed. Sujith play acted drinking arrack and getting drunk, and he tried to get arrack from neighbors, including one who obliged him until Sujith's grandmother intervened.

PAST-LIFE BEHAVIORS

Many of these children show behaviors that appear connected to the memories they report. Some show emotions toward the individual family members of the previous person that seem appropriate for the relationship the person had. A little girl may be deferential toward the previous husband or parents but bossy toward the younger siblings of the previous person, even though those individuals are now much older than she is. These emotions usually dissipate as the statements fade but not always. There is at least one case in which a little boy who talked about a past life eventually grew up to marry the widow of the previous person.[26]

Phobias are often a part of these cases.[27] In the unnatural death cases, about 35 percent of the children show intense fear toward the mode of death of the previous person. For instance, one little girl in Sri Lanka always hated being in water. When she was just an infant, three adults had to hold her down to give her a bath.

When she became old enough to talk, she described details from the life of a girl in a nearby village who had drowned.[28]

The children may show likes and dislikes that appear connected to previous lives. Addictive substances sometimes seem to carry their appeal into the next life, as some of these children will try to sneak cigarettes or even shots of liquor if the previous person was fond of them. Regarding dislikes, the clearest examples involve food in some of Stevenson's Burmese cases. He collected twenty-four examples of young children in Burma who said that they had been Japanese soldiers killed there during World War II. Some would complain about the spicy Burmese food and ask to eat raw fish instead.[29]

The children's play often shows signs of being connected to a previous life.[30] Some can play act the previous occupation for hours on end. One little boy said he had been a biscuit shopkeeper, and he played at being a shopkeeper compulsively, to the point that he would not go to school. He fell behind in his schoolwork, and his mother felt he was never really able to catch up.[31]

Another behavior in some of the cases is cross-gender behavior. Nine percent of the children report a life as a member of the opposite sex.[32] In those cases, the children often show gender nonconformity, and they sometimes identify with the opposite sex to the point that they appear transgender.[33] Although this sometimes lessens as the children age, some have shown continued gender dysphoria into adulthood.

The behavioral features in the cases raise the question of whether psychological factors may play a role in the phenomenon. Small testing studies have been conducted in Asia and the United States. In our evaluations of a group of fifteen American children, intelligence testing indicated above average IQ in both the verbal and quantitative domains. We also included the Child Behavior Checklist (CBCL), a common scale used in many clinics. The children we studied did not show any behavioral disturbances on the CBCL. The parents completed the Child Dissociative Checklist as well. Symptoms on the checklist include normal behaviors such as daydreaming, as well as more pathological symptoms suggesting a dissociative identity disorder (multiple personality disorder). The children scored very low on the checklist, meaning they demonstrated few signs of dissociation. Overall, our study indicated that these children are intelligent and do not exhibit any significant psychopathology.[34]

Haraldsson has published studies of two groups of children in Sri Lanka and one in Lebanon.[35] The children in his studies tended to show some mild behavioral issues (e.g., oppositional, obsessive, and perfectionistic traits), and Haraldsson speculated that some of them might be showing mild symptoms of a posttraumatic stress disorder (PTSD), related to events they reported from their past lives.[36] In general they demonstrated cognitive skills as good or better than controls, and performed well in school.

THE CASE OF JAMES LEININGER

James Leininger grew up as the son of a Christian couple in Louisiana.[37] When he was twenty-two months old, his father took him to a flight museum. James appeared transfixed and kept wanting to return to the World War II exhibit. They eventually left with a few toy planes, and when James played with them, he repeatedly crashed them into the family's coffee table, saying "Airplane crash on fire," producing dozens of scratches and dents in the table.

He began having nightmares a couple of months later. His mother would find him thrashing around and kicking his legs up in the air, screaming "Airplane crash on fire! Little man can't get out!" The nightmares occurred multiple times a week, and his parents, as well as his aunt, emphasized how disturbing they were to witness.

Several months into the nightmares, James began talking about them one night before going to sleep, saying that his plane had crashed on fire when it was shot by the Japanese. A couple of weeks later, when he was twenty-eight months old, his parents had another bedtime talk with him. He said his plane was a Corsair, a fighter plane that was developed during World War II. He talked a number of times about flying a Corsair, a plane that was not part of the exhibit at the flight museum. He also said he had flown off a boat named "Natoma." After the conversation, his father searched online for the word Natoma. After some effort, he found a description of USS *Natoma Bay*, an escort carrier stationed in the Pacific Ocean during World War II. He printed out the information and kept it, and with the footer showing the date the pages were printed, it serves as documentation that James gave the name Natoma when he was twenty-eight months old.

When his parents would ask who the little man in the plane was, James always answered "James," which his parents initially thought referred to his own name. His parents asked if anyone else was in the dream with James, and he gave the name Jack Larsen. James said he was a pilot, too.

When James was just over two and a half, his father was looking through a book on the battle of Iwo Jima, when James pointed to a picture showing an aerial view of the island and said that was where his plane was shot down. His father soon learned that *Natoma Bay* had taken part in the Iwo Jima operation and that a Jack Larsen had been a pilot on the ship.

Soon after his third birthday, James began drawing pictures. He drew battle scenes with ships and planes over and over again—his parents report he drew hundreds of them, suggesting possible posttraumatic play like that often seen in traumatized children. He signed them "James 3" and continued to do so after he turned four years old. He said he was the third James.

A couple of months after James turned four, he and his parents were interviewed by ABC News. The interview documents his parents' report of the details James had given at that point, including ones about how his plane had been shot down. A few months later, James's father attended his first *Natoma Bay* reunion. He learned that Jack Larsen was still alive, and they later met. He also learned that only one pilot from the ship had been killed in the Iwo Jima operation, a twenty-one-year-old man from Pennsylvania named James Huston, Jr. Since he was James, Jr., this could make James Leininger "the third James."

Here are James's statements and behaviors documented before Huston was identified, compared to details about Huston's life:

James Leininger	James Huston
Signed drawings "James 3"	Was James, Jr.
Flew off ship called Natoma	Pilot on *Natoma Bay*
Flew a Corsair	Had flown a Corsair
Shot down by the Japanese	Shot down by the Japanese
Died at Iwo Jima	The one *Natoma Bay* pilot killed in the Iwo Jima operation
"My airplane got shot in the engine, and it crashed in the water, and that's how I died."	Eyewitnesses reported Huston's plane was "hit head on, right on the middle of the engine."
Had nightmares of plane crashing into the water and sinking	Plane crashed into the water and quickly sank
Jack Larsen was there	Jack Larsen piloted the plane next to Huston's on the day Huston was killed

I interviewed James and his parents as James neared his twelfth birthday. His nightmares had ceased years before, and though he remained interested in planes, he no longer showed the obsessive play he had earlier. The memories themselves seemed to be gone by that age as well. I was able to observe James during both the formal interview and in informal settings. He presented as a typical boy his age and showed no evidence of any psychological disturbance. His parents remained Christians, but his father, who had initially been firmly opposed to the idea of reincarnation, joined James's mother in believing that James had remembered a past life.

THE CASE OF RYAN

Ryan was a five-year-old boy living with his parents in the Southwest United States when his mother contacted our office.[38] She reported that for the previous year, Ryan had been talking about going home to Hollywood. He would cry and plead for his mother to take him home so he could see his other family. One night, he said he wanted to tell his mother what it was like when you die. He began describing an awesome bright light and said you should go to the light. He said everyone comes back and that he knew her before. He claimed he picked her to be his mother.

Since Ryan seemed to be struggling with the memories, his mother went to their public library and picked up some books about Hollywood, in hopes that he would see landmarks that would help him better process his emotions. One of the books contained a picture from a 1932 movie called *Night After Night*. Ryan became excited when he saw it, and as he looked at the men in the photograph, he pointed at one of them and said, "Hey Mama, that's George. We did a picture together." He pointed at another and said, "And Mama, that guy's me. I found me." The photo showed six men, all looking at the two in the middle having a confrontation. Ryan's mother verified that the first one Ryan pointed to was George Raft, a film star in the

1930s and 1940s. The second one was off to one side. The book did not list the people in the picture, so his mother was unable to identify him. She later discovered that the man had no spoken lines in the movie, making him even harder to identify.

Ryan later told his mother, "I liked it better when I was big and I could go wherever and whenever I wanted to go. I hate being little." He also said, "I just can't live in these conditions. My last home was much better." He said, however, that the reason he had to come back was that he did not spend enough time with his family in his last life; he worked so much that he forgot that love was the most important thing.

Ryan made numerous statements about another life. He talked about dancing on stage in New York. He began doing a tap dance routine one day and said he had recalled it when he heard some music on a cartoon that sounded like the music he used to tap to. He said he had done it with two buddies. He had not taken formal lessons but had taught himself.

Ryan talked about traveling a lot. He said he had been to Paris and seen the Eiffel Tower. He also said he could not wait until he got big again and could "go on those big boats, wear fancy clothes, and dance with all the pretty ladies." He said that was how you saw the world, from a big boat. He mimicked dancing what appeared to be a waltz and said that doing the dip moves was his favorite part of dancing on a ship.

Another time, Ryan said he used to live somewhere with the word Rock or Mount in it, indicating it was part of a street address. He described living in a large house with an outdoor swimming pool. He also told his mother he had nightmares about a man called Senator Five, whom he described as the nastiest villain that ever lived. He said he had gone to New York to meet him. He also talked about working at an agency where people changed their names. At one point, he resisted going to kindergarten because after he would tell his class about Holly-wood and the agency in story time, the other children would make fun of him when he said the stories were real.

With the help of a Hollywood archival footage consultant, the second person in the picture Ryan had pointed to was posi-tively identified as a man named Marty Martyn. I then met with Ryan and his parents. I showed Ryan lineups of four pictures, with one picture being a person from Marty Martyn's life. Ryan was slow to cooperate and seemed to point at pictures randomly. Eventually, he became more serious and correctly identified Martyn's last wife and a picture of a young Martyn. He also picked out New York's senator during the late 1940s and 1950s, a Senator Ives whose name sounded very similar to the Senator Five that Ryan had mentioned. In addition, he correctly identi-fied Marty Martyn's name out of four that were read to him by his father, who did not know which name was correct.

Martyn's daughter was located. When Ryan learned about her, he seemed confused. People explained that she had grown up, but Ryan said he thought she was still little. He said he remembered her being not much bigger than he was then and that he really did not think he had been gone that long.

I met with Martyn's daughter, who was eight when her father died. Between her memory and the records that could be found, I was able to determine the basic facts of Martyn's life. It had seemed unlikely that a movie extra with no lines would have danced on Broadway, had a big house with a swimming pool, and traveled the world on big boats, but Martyn in fact did. After being born in Philadelphia, he and one of his sisters went to New York to be dancers. He danced in various revues on Broadway, and his sister became a well-known dancer there. He then moved to Los Angeles, having a life in Hollywood as Ryan had described. He began as an extra as well as a dance director.

He then became a Hollywood agent and set up the Marty Martyn Agency, a place where people changed their names as Ryan
said. He married four times, and he lived with his last wife in a
large home with a swimming pool. Ryan had said his address
had Rock or Mount in it, and Martyn's last home was on Roxbury. He and his wife went to Paris a lot, and they traveled to
Europe on the Queen Mary. His daughter confirmed that he
had indeed met Senator Ives from New York.

Some of the details Ryan gave did not match Marty Martyn's life. For example, he talked about a father who had raised
corn and then died. There were numerous other items that could
not be proved or disproved, since the sole sources of information
were public records and Martyn's daughter, who was only eight
when he died. Even so, over fifty of the details Ryan gave did fit
the man he pointed to in the picture.

Ryan met Martyn's daughter. He seemed intimidated by the
experience and said very little to her. He was also shown some of
the sites Martyn had known. The house on Roxbury had recently
been torn down, and a new one was being built in its place. Ryan
seemed okay with that. He enjoyed going to an apartment complex Marty had stayed in, as well as the Beverly Hills Hotel,
where Marty once had a bungalow. Following the trip, Ryan's
talk about Marty Martyn's life became infrequent unless something reminded him of it, and then stopped almost completely.
He told his mother it was time to just be a regular kid.

COLLECTION DATABASE

In recent years we have been coding each case in our files based
on two hundred variables and putting the coded information
into a database. This project, which involves over two thousand

cases, is ongoing. Nonetheless, we have already observed some interesting patterns.

We developed a strength-of-case scale that assigns points to cases based on four areas: birthmarks and birth defects, statements about the previous life made by the children that were verified to be accurate, behaviors of the children that appear related to the previous life, and the distance between the child's family and the family of the previous personality. We then applied the scale to the 799 cases that were in the database at the time the scale was created.[39]

The results showed that the apparent strength of cases did not correlate with the initial attitude that the children's parents had toward their child's statements. This means that parental enthusiasm or belief that their child was reincarnated did not make the cases appear stronger than they actually were. The strength-of-case score did correlate with the amount of acceptance of the children's claims by the deceased individuals' families, suggesting that those families used criteria similar to the ones in the scale in assessing the situation. The strength of the cases was inversely correlated with the age that the children began talking about the previous life; that is, the younger the child was when he started reporting memories, the stronger the case tended to be. The strength was positively related to the amount of emotion the children showed when discussing the memories, along with the amount of facial resemblance between the children and the deceased individuals. These findings were consistent with the idea that in the stronger cases, more past-life residue had carried over to the new life.

We also used the database to analyze memories that some children report of events between the death of the previous personality and the birth of the child.[40] Approximately 20 percent of the children in our cases make such reports, some describing terrestrial events such as the previous person's funeral and some reporting activities in other realms. The analysis showed that the children who report interval memories make more statements about the previous life that are verified to be accurate than the children who do not. They recall more names from the previous life, have higher scores on the strength-of-case scale, and are more likely to state the names of the deceased individuals and to give accurate details about their deaths. Thus, the interval memories seem to accompany stronger memories of a previous life.

When we looked closely at thirty-five such cases in Burma, we saw that the interval memories could be broken down into three parts: a transitional stage, a stable stage in one location, and a return stage involving conception or choosing parents. We compared the reports of interval memories by the Burmese children to published reports of near-death experiences (NDEs). We found similarities in the interval reports to the transcendental aspects of NDEs and significant areas of overlap with Asian NDEs in particular. This suggests that we can think of interval memories and NDEs as examples of the same overall phenomenon: reports of an afterlife.

We have also explored issues related to the death of the previous person. An unnatural death is a prominent part of many of these cases. It also seems that dying young is important, separate from the issue of suffering an unnatural death. Using the database, we can separate out the unnatural cases and look just at the cases involving natural deaths. When we do that, we find that even when the previous people died naturally, a quarter of them were under the age of fifteen, and, in general, the younger the age group we consider, the more cases we have. There seems to be something about a life that ends in an unnatural way or ends early that makes it more likely that a child will later talk about it.

Several features also suggest that if these are indeed cases of reincarnation, characteristics of the end of the past life affect where the child is reborn—with unusual deaths more likely to lead to a new life with strangers, while ordinary deaths more often produce a return to the same family. In the stranger cases, the previous person tends to die younger, to die by unnatural means, or to die more unexpectedly, even when the death was due to natural causes. One interpretation of these findings is that in the stranger cases, factors such as a premature or violent death cause the individual to come back to another life here. In the same-family cases, on the other hand, perhaps it is a strong emotional pull that leads the individual to return, in that case back to their previous family.

POSSIBLE CONVENTIONAL
EXPLANATIONS FOR THE CASES

In attempting to account for how very young children can accurately describe the life of a deceased person otherwise unknown to them, various explanations have been considered. One conventional explanation that is possible for many of the cases is that faulty recollections by the children's families cause them to believe that their children gave more details about the previous life than they actually did. Researchers have not investigated most of the cases until after the deceased individual has been identified. Unless somebody recorded the child's statements early on, the possibility exists that after a child makes a few general statements about a previous life, the parents find another family that has lost a family member, and once the two families exchange information, they come to believe falsely that the child had demonstrated specific knowledge about the previous life beforehand, when in fact they did not.[41]

Two studies have looked at this possibility. Keil reinvestigated fifteen cases that Stevenson had studied twenty years earlier to see if the reports by the families had become exaggerated over time.[42] He and Stevenson found that the family told a stronger story during the second investigation in only one case, when they described an incident they had not reported before. In three other cases, the strength of the reports remained unchanged. Some of the details were different in one report compared to the other, but overall, the cases had not grown stronger or weaker over time. The reports of the other eleven cases had become weaker by the time that Keil talked with the families. This was often because the families gave fewer details than they had given Stevenson years before. Thus, the study showed that the cases had not grown stronger in people's minds over time; in fact, they had become weaker as families, rather than creating new statements in their minds, had remembered fewer of them.

The other study involved statistical work comparing two groups of cases from India and Sri Lanka—one group in which written records were made of the children's statements before the family of the child met the previous person's family, and the other group for which no such records were made.[43] The percentage of correct statements was essentially identical in both groups (77 percent versus 78 percent), but the cases with written documentation included a larger number of statements on average than the ones without written records (twenty-six versus nineteen). The findings therefore contradict the idea that families later exaggerate the amount of knowledge that the children had conveyed about the previous life before the families met.

Taken together, these two studies cast doubt on what appears to be the best way to explain many of the cases with a conventional mechanism—that witnesses incorrectly remember the children's statements about the previous life as being more impressive than they actually were.

In summary, the strongest cases we have studied provide evidence that memories, emotions, and evidence of physical trauma may, at least under certain circumstances, carry over from one life to another one. If such carryover does occur, it is not necessarily universal or even common, since the experiences of the children with apparent memories of previous lives may be exceptions rather than the norm. Nonetheless, the cases contribute to the body of empirical evidence suggesting that some aspects of mind may in some way transcend the physical body as currently understood, and perhaps even survive bodily death.

4

MEDIUMS, APPARITIONS, AND DEATHBED EXPERIENCES

EMILY WILLIAMS KELLY

ollowing Darwin's revolutionary transformation of the landscape of biology in the mid-nineteenth century, the notion that all aspects of life—including the nature of mind—would be explicable in terms of straightforward physical mechanisms subject to the processes of evolution became increasingly popular. However, also during the last decades of the nineteenth century, a remarkable group of primarily British scientists and scholars thought that the connection between mind and the physical body might not be so straightforward. This group was interested in the scientific investigation of phenomena suggesting that under certain conditions some aspect of mind might transcend the body—perhaps even surviving death of the body. They strongly believed that the methods of science, observational as well as experimental, could be used to study human experience in all its variety. In 1882, members of this group formed an organization—the Society for Psychical Research (SPR)—to support the investigation of questions at the frontiers of the scientific understanding of the human mind.[1]

Over the course of the next few decades, those involved in psychical research studied and published a vast amount of material—experiences reported to them by ordinary persons, as well as experiments that they themselves conducted—relevant to the question of whether aspects of human personality survive death and, more generally, whether mind is wholly dependent on the body. In the first two decades alone, their books and papers amounted to tens of thousands of pages, including the classic volumes *Phantasms of the Living*[2] and *Human Personality and Its Survival of Bodily Death*,[3] as well as twenty-two large volumes of the *Proceedings of the Society for Psychical Research* and eleven volumes of the *Journal of the Society for Psychical Research*.[4] Few people know that this empirically based literature relating to the question of survival after death even exists, let alone of its scope and high quality.

For fifty years, our group at the University of Virginia has been the only university-based research group devoted primarily to the scientific study of evidence for survival after death. Near-death experiences (NDEs) and cases of the reincarnation type (CORT) have been discussed in chapters 2 and 3. In this chapter, I will outline three other kinds of evidence that I think are particularly important. But before beginning, I wish to emphasize two crucial points for readers to keep in mind. First, the evidence for survival does not depend on one case, or even on one line of research. There are multiple lines of research, and there are scores, hundreds, and even thousands of experiences that have been reported for each area of research. Particular cases and particular lines of research have their strengths and weaknesses, but the important point is that they all seem to be converging on one main idea: there is evidence that, in some form, certain aspects of one's mind or personality may survive death of the physical body.

In addition to the sheer quantity of the evidence, I also wish to emphasize its quality. It is not appropriate to collectively dismiss thousands of published experiences as mere hearsay or anecdotes. Many of them are firsthand accounts that have been painstakingly investigated, often with relevant documents and corroboration by witnesses to support them. In this chapter, I have chosen to illustrate the various kinds of experiences with contemporary examples reported to us at the University of Virginia, partly because these can be described more succinctly than some of the older long case reports and partly because I wish to emphasize that such experiences were not confined to earlier times, but are still happening—frequently—today. The examples I give here have usually not been investigated in depth, but they are startlingly similar to the older, carefully investigated cases.

In short—if one is to take a position on the question of survival after death, whether as a skeptic or a believer, that opinion should be based on an awareness of the breadth and depth of the evidence available, and not on uninformed assumptions or one or two isolated examples.

MEDIUMSHIP

The first area I will discuss is what is known as mediumship. Mediums are people who (usually) go into a trance or some other altered state and, either through speaking or automatic (unconscious) writing, seem to be communicating with actual, identifiable deceased people. The study of mediums—much of it focused on several especially talented mediums—was a major area of research in the latter part of the nineteenth century and the first part of the twentieth century. A great deal of high-quality material was published, showing that these mediums had specific knowledge about specific deceased people that they could not have learned in any normal way. Everyone adequately knowledgeable about mediumship agreed on this latter point, but they disagreed on the interpretation. Some people thought that telepathy between the medium and living people could account for the source of the material. Other people thought it was what it appeared to be on the surface—communications from deceased people.[5]

In the face of this impasse over how to interpret these cases, researchers began focusing on cases in which telepathy seemed less likely. In an ordinary "sitting" (or "seance"), a person (called the "sitter") was brought by researchers to the medium, often anonymously, for a face-to-face meeting with the medium. In these situations, the medium usually gave information (called a "reading") that the sitter recognized as relevant to a deceased loved one. Sometimes, however, a person whom the sitter did not recognize seemed to be communicating. In some of these cases, the unknown person (called a "drop in") gave enough information to allow investigators to identify him or her.[6] Since no one present with the medium knew this person, telepathy between the medium and the sitters is not a likely explanation. Moreover, the motivation to communicate seems to have come from the deceased person, not living people. Unfortunately, we have to take drop-in cases when they present themselves; so far no one has been able to induce one deliberately.

A more promising line of research, therefore, has been what is known as "proxy" sittings. In these sittings, the investigator or some other stranger who does not know the particular deceased person in question sits with the medium as a proxy for the real sitter. The advantage of such proxy sittings is not only that it weakens the explanation of telepathy from the sitter, but it also weakens the two "normal" explanations that may apply to ordinary, uncontrolled sittings with mediums: that mediums get feedback from the sitters that allows them to elaborate and produce correct statements; or that what the mediums said was so vague or general that by chance it could apply to many people, but be falsely interpreted by grieving sitters as specifically applying to them.

Many such proxy studies were conducted during the early years of the twentieth century.[7] However, it can be argued that even proxy and drop-in cases are not convincing evidence that mediumship is still not explicable by some kind of "super extra-sensory perception" (super-ESP or super-psi) between living people. The debate on how to interpret these phenomena is an important one to have, but one unfortunate result of this disagreement in interpretation was to discourage serious research on mediumship. We may never resolve the debate, but if we are ever to do so, the research must continue.

In recent years, renewed interest in mediumship among the public has been sparked by television shows, movies, and books by and about mediums. A few years ago, therefore, I decided to do a study with several mediums, in the hope of reviving and encouraging research of the caliber of what was done a century ago. I particularly hoped to identify mediums who can do well under the controlled conditions of a "proxy" sitting. Briefly, the study was as follows: nine mediums and forty sitters participated. The mediums were sent photographs of deceased people. These photographs were "neutral," in the sense that they showed the deceased person alone and not engaged in activities that might provide significant information to the medium. The mediums were given no other information. Two of the mediums each did readings for six deceased people, and the other seven mediums each did readings for four deceased people. None of the sitters were present; I or a co-worker served as the proxy and audio-recorded the reading. Each of the forty sitters was sent six transcribed readings—the one intended for their deceased loved one, as well as five others intended for five other deceased people of the same gender and general age. Sitters were asked to pick the correct one from the group. They were able to do so at a rate far greater than chance, with fourteen of the thirty-eight people who returned their choices picking the correct reading out of the group of six.

The methodological and statistical details are published in the primary scientific literature.[8] Here I want to concentrate instead on providing some examples of the kinds of things the mediums said that led the sitters to pick the correct reading.

1. In one reading, the medium said: "I feel like the hair I see here [in the photo that the medium was looking at] is gone, so I have to go with cancer or something that would take the hair away." Then later she said: "Her hair—at some point she's kind of teasing [it], she tried many colors. I think she experimented with color a lot before her passing." The medium also said: "I feel I'm up in Northampton, Massachusetts. . . . Northampton does have that kind of college town beatnik kind of feel to it."

The girl's mother confirmed that her daughter had died of cancer, had dyed her hair "hot pink" before her surgery, and had later shaved her head when her hair began falling out. Importantly, her hair looked normal in the photo that was shown to the medium. Also, although the girl lived and died in Texas, her mother said that "this [Smith College in Northampton, Massachusetts] is where she told a friend she wanted to go to college."

2. In another reading, the medium referred to "a lady that is very much, was influential in his [the deceased person's] formative years. So, whether that is mother or whether that is grandmother. . . . She can strangle a chicken."

The sitter commented that her grandmother (who was the deceased person's mother) "killed chickens. It freaked me out the first time I saw her do this. I cried so hard that my parents had to take me home. So the chicken strangling is a big deal. . . . In fact I often referred to my sweet grandmother as the chicken killer."

In the same sitting, the medium referred to "Mike, Mikey, Michael." The sitter's brother (son of the deceased person) was known as "Mikey" when young, "Michael" as he grew older, and finally "Mike."

3. In another example, among many other details the sitter commented especially on the medium's statement: "He said I don't know why they keep that clock if they are not going to make it work. So somebody connected directly to him has a clock that either is not wound up, or they let it run down, or it's standing there just quiet. And he said what's the point in having a clock that isn't running? So, somebody should know about that and [it] should give them quite laughter [sic]." The sitter did laugh over this, because a grandfather clock that her husband had kept wound had not been wound since his death. The medium had also commented that "he can be on a soap box, hammering it"; his children when young had frequently complained about "Dad being on his soap box."

This study does not advance the survival/super-psi debate over previous proxy studies. What it did do, however, was show that a revival of research of the quality and scope of that done a century ago may yet be possible, because there are still mediums who can do well under controlled conditions. Not all mediums, even very good ones, will be successful in proxy sittings. These are very artificial conditions, different from the way mediums usually do things. But if we can find mediums able and willing to participate in controlled, and particularly proxy, sittings, such research may help advance our understanding of mediumship and its relevance to the question of survival.

APPARITIONS

Another important area of research is the study of apparitions, particularly what are known as "crisis" apparitions. These are experiences in which a person sees an actual apparition, hears a voice, has a dream or intuition, or feels the presence of another person, and the experience coincides with the time that the other person, at a distance, is undergoing some sort of crisis, usually death. Like mediumship, this was a huge area of study for the early psychical researchers, and they investigated and published literally thousands of these cases in the late nineteenth and early twentieth centuries. But this is another area where the research largely died away. With the rise of experimental parapsychology in the 1930s and 1940s, especially in the United States, the investigation of spontaneous experiences such as apparitions diminished—not going away entirely, but abandoned for the most part for experimental studies of ESP (chapter 5).

Our research group at the University of Virginia has been virtually the only one focusing primarily on spontaneous experiences. Over the years many people have written to tell us about experiences that they have had, and people today are reporting exactly the kinds of experiences that people were reporting over one hundred years ago. We just need to look for them and, by making such experiences better known, reduce the reluctance and fear many people have to tell others what they have experienced. As one person said when she wrote to us: "I've never told anyone else [about my experiences]. I thought everyone would think I was crazy. I hope you don't."

One of the most common experiences reported to us are crisis cases, whether a dream, an intuition, or an actual apparition. I will present here a few that are typical, just to give a sense of what they are like.

4. "My father died of [a heart attack] when I was 10 years old. This was quite unexpected . . . He was 42 years old. I told my teacher that my father had died and shortly thereafter, a close friend of my mother arrived [at school] to pick me up."

The child had no normal way of knowing that her father had died. There was apparently no apparition or any other sensory experience; she just suddenly had this intuition that her father had died. We do not know what state of mind she was in when she had her experience, but many crisis experiences occur in a dream or when the person is in some other kind of altered state of consciousness.

5. "I had a dream that my grandmother was calling me from her room. The next thing I knew I was talking to her about her taking a trip. In the morning or sometime that afternoon I learned she had died the night before. I wasn't surprised or shocked; it was as if she had already told me she was going."

One interesting note about this case is that several years later this woman had an NDE when she had severe complications during childbirth. In the NDE this same deceased grandmother appeared to her and sent her back, a circumstance known to happen in some NDEs (chapter 2).

In the next case it was unclear whether the experience was a dream, a waking sensory experience, or both. It was reported by a man who had been very close to his cousin, Bob:

6. "I was in bed one night, and I'm not sure whether I dreamt this or just what happened, but I could definitely hear [Bob] knocking on my window and on my door and asking to be let in, and telling me he needed help. It was just so realistic that I actually got up out of bed and went to the front door to see if he was there. Nobody was there, of course, and I went back to bed, and I heard it again. Needless to say, I didn't sleep much more that night. . . .

I got up at my usual time and went to my shop and shortly after I got into my shop my phone rang, and it was Bob's sister's son calling to tell me that he [Bob] committed suicide during the night."

A particularly important aspect of this experience is that the man had told his wife what happened before they learned about Bob's death. They both had woken up in the night. She did not hear anything, but he told her what he had experienced, and she confirmed to me that he had told her this. The early psychical researchers had emphasized getting such corroboration whenever possible. Getting corroboration is not meant to imply that the investigator does not believe the person, but it is helpful to have to address the possibility of inaccurate memories constructed after the fact. A critic of such a report might say, for example: "Oh, he later believed that he had this dream the night of the death, but he really didn't. It happened the next night, when he was upset, and he is simply recalling the time sequence erroneously." Corroboration from a second person helps address this kind of argument.

In the next case the person clearly was awake when he had the experience:

7. "I saw my grandfather walking down the hall of my house, which is 1200 miles [away] from his in Kentucky. He died of a massive coronary. When I saw him, he was almost opaque in appearance, as if you were viewing a hologram. I received a phone call 2 days later that he had passed. The time I saw him almost precisely coincides with the time of his death, [but] I was not told of his death for 2 days. I was extremely close to him, and the family frankly was trying to figure out which one of them should tell me."

As he said, this man was "extremely close" to his grandfather. A strong emotional bond might seem to increase the likelihood of such an apparitional experience at the time of death.

However, there are other situations when it is puzzling why the experience happened to the person that it did:

8. "When I was 18 years old and still living at my parents' . . . I was awakened at 1 AM. I opened my eyes and saw my boyfriend's brother-in-law standing at the foot of my bed. I don't know why, but I felt scared. I looked into his eyes and screamed. My parents ran into my room, turned on the light, and sat on my bed. I told them what happened, and they said it was a dream, so go back to bed. Just as they were walking out of my room, the telephone rang. My mom answered the phone, and said it was my boyfriend, he apologized to her for calling that late, but just wanted to tell us that his sister found her husband dead in their apartment when she arrived home from work around 11 that night."

There did not seem to be a close emotional bond here. Why did the experience not happen to the man's wife, or even his brother-in-law? Was the percipient a particularly receptive person in some unknown way, and others closer to the deceased man were not? Or is there some other factor, or combination of factors? The more experiences that we learn about, the more we realize that we really understand very little about why, when, and to whom they happen.

There is also another kind of crisis case. There is a folk song about a clock stopping when its owner died.[9] Many experiences have been reported to us, not of clocks stopping, but of some other kind of physical event that happened at the time of a death.

9. "A year and a half ago, my mother-in-law passed on unexpectedly. The night she died, my husband, 2 children, and myself were sitting on the couch watching a movie. All of a sudden I felt a cold breeze, like someone cold walked past quickly toward the kitchen, and then the picture of the Last Supper, which was

a gift from my mother-in-law, fell off the wall in the kitchen. We all felt the cold, but my husband said it was just a breeze and the windows are drafty. He said the picture falling was a coincidence. That night at 3 AM, his sister called us crying and saying their mom died of a heart attack."

Objects moving and cold breezes are relatively frequently reported, but other experiences are much more unusual:

10. "The day my Grandma died I was headed out to work about 3 AM. I knew she was very sick, and I had planned to fly to New Jersey later that day to see her because I knew time was short. As I went to put the key in the car, I felt this overwhelming presence, and all of a sudden my driver window shattered, and I thought to myself Grandma just died. As I got to work, I could not shake the feeling of grandma right next to me. The phone call came; Grandma died at the same time I felt her and the window shattered. I felt this was Grandma's way of telling me she is gone."

How often does a car window shatter like this? It may happen occasionally when it is forty degrees below zero in Maine. But this woman was in Florida. Clearly we need a better explanation than chance coincidence.

The importance of these crisis cases is that they are something more than subjective hallucinations, since the person having the experience did not yet know the other person had died. Indeed, the experiences seem to many people to be a kind of communication from the deceased person, as if the person had come to announce his or her death. But there are many similar experiences that do not occur in crisis situations, but after the person has died and the person seeing the apparition knows about the death. These postdeath experiences might be hallucinations caused by grief or wishful thinking. But because they are so similar to the crisis ones, perhaps we should not be too hasty in dismissing them simply as imagination, and should look more seriously at the possibility that the same process—whatever that is—is going on in both situations.

Here is a typical example, reported to us:

11. "I saw an older lady looking at my newborn daughter in her crib when I got up to use the bathroom. It was my grandmother who passed away while I was pregnant."

This might have been a hallucination resulting from the granddaughter's grief because her grandmother had not lived long enough to see the baby. But it also seems possible that the great-grandmother wanted to come back and see her new great-grandchild, especially in light of similar "crisis" cases in which grief could not yet have been a factor.

In this case, the woman actually saw her grandmother. In other cases, there is just a very strong feeling of presence, that people often find hard to describe. There is no clear sensory component to it, just a feeling that someone—even a particular someone—is there. Experiences of a sense of presence of a deceased person also occur in connection with NDEs (chapter 2). One conjecture by some of the early psychical researchers was that the sense of presence might be a kind of incipient, or beginning-stage, apparition that does not develop far enough to become an actual full-blown sensory experience. Here is an example:

12. "I was simply driving home one evening when I had the distinct feeling that my deceased father was in the car with me. It wasn't a creepy feeling, like a ghost was there; it was just a very real feeling like he was still alive and riding in my passenger's seat . . . just like we were going out to pick up a pizza or

something. I knew it couldn't be real, but I had to fight myself not to turn my head and look to see if he was actually sitting there. I could sense that he was laughing about something too, and I wondered if he was a spirit being who could read my thoughts and [was] amused by my unbelief."

Related to experiences of a feeling of presence, but taking on a sensory aspect, are smells. Many people report smelling either a familiar aftershave, cologne, perfume, cigar, whiskey, flowers—some smell that they associated with the deceased person. Smells might seem to be a completely subjective hallucination, but there are cases in which more than one person smells it, suggesting something more objective in origin:

13. "My mother-in law passed on in '99. After she had died, my husband and I had returned home. As we walked up the walkway to our front door, there was a sudden burst of fragrance. Very overwhelming and strong. My husband and I looked at each other, and both commented that it smelled like his mother's cologne. It seemed to come out of nowhere and was very strong for about 10–20 seconds, then it was gone."

I mentioned earlier that some crisis experiences involve physical events, such as the case with the shattering of the car window. There are also experiences like this among noncrisis cases. Here is one that also involved glass, although in this case the glass did not break:

14. "A friend of mine and I were talking about my deceased son . . . when all at once four glasses, hanging on hooks under the cabinet, started to move. They moved so hard I thought they would break. After five minutes they stopped moving. When my daughter came over that evening, I was telling her about it, and they started moving and hitting against one another all over

again. . . . Those glasses have moved for no reason several times when we discuss my deceased son."

We have reports in which the deceased person seemed to be warning somebody about something.

15. "I was driving home from work and was in a big hurry. . . . I was about to change lanes when I heard, as plainly as day, my father's voice come from the back seat, yelling 'Butchie, look out.' No other person ever called me Butchie, and my father had been dead for 3 years. When I immediately stayed in my lane, a pick-up truck came around on the wrong side. Had I changed lanes, I am certain I would have had a very serious accident."

One particularly important type of apparitional experience is that known as a "reciprocal" apparition. In such instances, one person seems to have an "out-of-body" experience and "visit" a distant location, while people at that location feel a sense of presence or see an apparition of that person. Such experiences are quite rare, but at least thirty have been reported.[10] One such experience was reported to us recently by a nurse:

16. "I became very close to a quadriplegic man—my age—who was admitted several times for pneumonia. I became close to his family as well. He had been a quad since 19 years old. His family was incredible, and he actually was one of the more clever and witty persons I have ever known. He was complicated to care for and required diligence and very special treatments to recover. I was very protective of him.

"One particular time he was ill and was not getting better in the facility where I worked. I helped contact a specialist, and he was transferred to the large teaching hospital where I worked previously and to a physician I knew well.

"I was trying to respect the professional relationship—I did not want to intrude—I spoke with the sister several times who called me and told me how he was doing. I was in school again and working and kept missing opportunities to see him.

"One night after a long day I was very tired and really he was on my mind. I was concerned that he was doing poorly and was feeling very guilty that I had not visited him. I knew he would be well cared for, but since he had several complications and dysreflexia, I worried whether he would survive this bout. I remember I was teary before I crawled into bed and went to sleep with him on my mind.

"I dreamed I was in the unit where he was. I dreamed I was at the end of the bed and that I was talking to him and telling him to keep fighting and not give up hope. I awoke still thinking of him.

"His sister called very soon after and said his fever had broken and he was better. He had told her that I was with him one night in the middle of the night and was telling him how to get better and that he needed to keep fighting. He said that I was standing at the end of his bed, it was the middle of the night and he wondered why I was there so late (I did not work in this facility).

"I was shocked. My dream had been so real, and I felt like I saw him clearly in the dream. I don't remember what I said to his sister. I felt really strange but recognized that I went—somehow—to him."

The research on apparitions and related phenomena has revealed one important characteristic: apparitions often display what might be called "quasi-physical" properties. On the one hand, they are clearly not wholly subjective hallucinations. In some way they evoke sensory perceptions—including sight, hearing, smell, and touch; more importantly, in many instances these sensory perceptions are shared. More than one person present may see the apparition from their differing perspectives, hear the same voice, or smell the same scent (as in #13 described above). Even more intriguing are cases in which the apparition is reflected in a mirror or blocks the light of a lamp when it passes in front of it.[11] The "reciprocal" apparitions I described above similarly suggest that some aspect of a person is sufficiently "physical" to have stimulated a sensory perception in other people at a distant location.

On the other hand, apparitions are clearly not physical in the ordinary sense. They are opaque or transparent to varying degrees, ranging from a felt sense of presence only to a completely realistic-appearing figure. They usually appear and disappear suddenly, perhaps seeming to pass through closed doors. And sometimes not all people present see the apparition. They seem to be something between matter and mind as we currently conceive them, and as such they may prove especially important as we seek to understand the relationship between mental and physical phenomena.

DEATHBED VISIONS

Another line of research I will mention, somewhat related to apparition cases, are what we call deathbed visions. Unlike NDEs, in which the person is close to death but then recovers, these are experiences of people who actually go on to die soon afterward. In a typical experience, the dying person seems to be seeing or talking with someone who is not physically in the room, usually a deceased spouse, parent, friend, or sibling. Not much research has yet been done in this area, although some collections of these cases were published early in the twentieth century,[12] and some preliminary research interviewing doctors and nurses about deathbed visions was done in the 1960s and 1970s.[13] Recently, a few people, particularly in the United Kingdom, have begun research in this area,[14] and we too hope to work with local hospice or palliative care people to learn more, not only about these experiences but also about experiences their family members may have at or near the time of death. We have talked with hospice people who say such experiences are extremely common, and some even say that they know death is near when a patient starts talking to a deceased person. Here is a typical example, reported to us:

17. "Right before my grandma died she looked up with a beautiful smile and said 'Hi, Edward [this was my grandpa]. I have missed you so much.' Then she died."

Because systematic research in this area is just beginning, we cannot yet answer even basic questions. How common are these experiences? What kinds of things are people actually reporting? When and to whom do they happen? Preliminary studies suggest that they seem less likely to happen when people have been given sedative drugs. In contrast to producing the experiences, sedating drugs seem to dampen them. Interpretations of these experiences are also complicated by cases in which, not the dying person, but a bystander at the bedside saw something unusual. For example:

18. "When my father died, I saw what I feel was his soul leave his body. It was like a shimmer, like when you look at the road on a hot day, as close as I can describe it."

In the following case reported to us,[15] the dying person had suffered a stroke and was in a coma, but her son-in-law (a physician) had this experience:

19. "I was standing by her bed and no one else was in the room. She had an agonal inspiration, and at that moment I had a very

clear picture of [her deceased husband] standing across from me with his arms outstretched, and he said: '[Her nickname], I've been waiting for you'."

Incidentally, the percipient in this case was my own uncle; and interestingly, I had never heard about his experience, in all my years of growing up, until I went into this field of research, when my uncle then told me about it. Many more such experiences likely remain unknown to others, even to close relatives and friends of the ones who have had such experiences.

People often assume that deathbed visions are produced by expectation or wishful thinking on the part of the dying person. But there are many experiences in which a dying person sees someone they did not know was dead.[16] Ian Stevenson learned about the following case in the 1950s, when communication was not as rapid as it is today:

20. "An elderly woman . . . became seriously ill. When the doctors said that she did not have long to live, the family gathered around her bed. Suddenly she seemed much more alert and the expression on her face changed to one of great pleasure and excitement. She raised herself slightly and said: 'Oh, Will, are you there?'—and fell back dead."

No one present was named Will, but shortly afterward they learned that the woman's brother, Will, who lived in England, had died two days before her own death.

DEATHBED LUCIDITY

There is another type of experience that we think is an especially important type of deathbed experience. We call them terminal lucidity, or revival, cases, because they often involve people who are comatose or suffering from dementia, and who seem to regain their faculties shortly before death.[17] Cases like this have been mentioned in the medical literature for centuries. For example, the great eighteenth-century physician Benjamin Rush referred to them briefly in his medical textbook, as if they were common knowledge, saying "most of mad people discover a greater or less degree of reason in the last days or hours of their lives."[18] These days one does not often hear about them, and so far there has been no research specifically in this area.

Yet there was one example reported in a recent issue of *Time*.[19] This particular issue of *Time* was focused on the mind-body problem, and written almost exclusively by scientists who had the view that everything about consciousness will ultimately be explained as knowledge about the brain grows. And then there was this beautiful little essay by an orthopedic surgeon, Scott Haig, tacked on at the very end of the issue. I summarize what it said:

21. David was dying of lung cancer that had spread to his brain. A brain scan showed extensive damage and in the days before his death he was completely unresponsive. When Dr. Haig left one evening, David was clearly dying, but the following morning a nurse reported: "He woke up, you know, doctor—just after you left—and said goodbye to all of them. . . . He talked to them and patted them and smiled for about five minutes. Then he went out again, and he passed in the hour."

Dr. Haig's comment: "It wasn't David's brain that woke him up to say goodbye that Friday. His brain had already been destroyed. . . . What woke my patient that Friday was simply his mind, forcing its way through a broken brain, a father's final act to comfort his family."

A similar case, this one involving not brain cancer but dementia, was reported to me recently:

22. "My grandmother chatted with me the week before she died, after not talking or reacting for a number of years—it was like talking to Rip van Winkle, as we chatted about what had been happening with family members while she was 'gone'."

Even though these revival experiences, unlike apparitions or deathbed visions, do not involve deceased people and so may not seem directly relevant to the survival question, they are in fact extremely important indirect evidence for survival. One of the biggest problems most people have accepting survival-related evidence is that we have so much other evidence, not only from science but from our own daily observations, of the undeniable correlation between mind and brain. But some people, including the psychologist and philosopher William James,[20] have understood that the correlation of mind and brain does not automatically lead to the conclusion that brain causes or produces consciousness. The correlation may be looked at in a different way. It might instead mean that the brain filters or restricts consciousness to allow efficient functioning in the physical world. If we were exposed most of the time to an unfiltered consciousness, we might be too overwhelmed to function. In this view, the brain functions to restrict and channel our consciousness, rather than produce it.[21]

Like many NDEs, revival cases seem to support this interpretation: when the brain is shutting down, its hold on the mind loosens, and the mind is then freed to function more fully. As Dr. Haig said about his patient, his mind broke free from its diseased brain and could function again. I mentioned earlier that many of the experiences we study—from having a telepathic experience, to seeing an apparition, to communicating with a deceased person while in a mediumistic trance, to having a deathbed vision or an NDE—happen when a person is in an altered state of some kind, that is, when the normal relationship between the brain and the mind has shifted or changed in some way. If we can loosen the grip that the brain ordinarily has on the mind, then we might free the mind to do things we cannot ordinarily do. Perhaps death provides the ultimate loosening.

5

PARANORMAL PHENOMENA, THE *SIDDHIS*, AND AN EMERGING PATH TOWARD RECONCILIATION OF SCIENCE AND SPIRITUALITY

EDWARD F. KELLY

In this chapter I will begin by discussing paranormal phenomena and psychic abilities, including their relationship to the "*siddhis*"—"perfections" or "attainments" as described in the vast literature on yoga and meditation. In the Western world we speak of such things under the general heading of what are often called "psi" phenomena, where the word "psi" is a theory-neutral term used to denote the relevant class of phenomena while making no unwarranted claims or suggestions as to their underlying mechanisms. The core feature defining such phenomena is that information appears to flow between an organism such as a human being and some part or aspect of the external world despite the presence of a *barrier*, such as physical shielding or separation in space and/or time, which according to current physical science should be sufficient to prevent that flow from happening. In a crisis-apparition case, for example, the percipient somehow directly acquires veridical and often detailed information about the mortal injury or death of a distant loved one, somehow overcoming the barrier of spatial separation. It is precisely this violation of expectations based on current physicalist understanding that makes psi phenomena so important theoretically.

There are two main classes of psi events, corresponding to the direction of information flow. Occurrences in which information flows *into* a person are identified by terms such as "telepathy" (direct or unmediated knowledge of contents of another person's mind), "clairvoyance" (direct or unmediated knowledge of some spatially remote physical state of affairs), "precognition" (direct or noninferential knowledge of future events), and "retrocognition" (direct or unmediated knowledge of past events), all of which are hard-to-distinguish subtypes of the more general category of "extrasensory perception" (ESP). The other direction of flow across a physical barrier involves direct production of physical effects on the outside world without use of one's motor apparatus or other normal physical means, however complicated. Occurrences of this sort fall under the heading of "psychokinesis" (PK) or "mind-over-matter" phenomena.

The term "*siddhi*" appears throughout the vast literature of Eastern spiritual traditions, and is used to denote an immense variety of unusual capacities thought to be acquired through means such as psychoactive plants or fungi, magical rituals and, especially, intense practice of meditation in various forms. *Siddhi* is here understood as the more inclusive term, and paranormal phenomena as described above constitute only a small subset, although an especially important one theoretically as indicated above.

It was mentioned in chapter 4 that the ongoing discussion about postmortem survival revolves in large part around issues as to whether particular forms of ostensible evidence for survival really demonstrate the continued existence in some form of deceased persons, or whether that evidence can instead always be interpreted in terms of psi-type interactions involving only living persons—the "survival versus super-psi" debate, as it has often been called. That debate now has a scientific history dating back more than 130 years, and it rages on with no clear, unequivocal answer, and with people more or less equally knowledgeable about the relevant literature lining up on both sides.[1] These are very difficult issues, to be sure, but I myself have recently come to believe, for reasons I will indicate later, that the cumulative evidence, although not compelling, is sufficient to justify rational belief in the possibility of some form of survival. A world that contains such a possibility is obviously radically different, and in humanly significant ways, from the world envisioned by contemporary mainstream science.

Psi phenomena thus lie at the heart of the survival debate, and either horn of this interpretive dilemma—living-agent psi or survival—conflicts inescapably with current mainstream scientific orthodoxy. That is precisely what makes psi so threatening to many mainstream scientists and yet so interesting to us. If the standard "production" model of brain-mind relations is correct, then these things simply could not happen. The fact that they do happen means that something is deeply wrong with that model, and we have got to enlarge our current scientific framework in some way that will enable us to accommodate them. An exciting challenge, and one to which I will return below.

Psi phenomena are not only problematic for contemporary mainstream science but ironically they are also often marginalized or ignored by the world's religious traditions, in which they are widely reported to occur. Particularly on the non-Tantric side of the Hindu tradition, for example, there is a very general tendency, despite universal agreement as to the existence of the phenomena, to discourage interest in or involvement with them. Patanjali says this explicitly in Book 3 of the *Yoga Sutras*, right in the middle of his catalogue of *siddhis*, when he declares that these are "powers in the worldly state, but obstacles to *Samadhi*."[2] In other words, they are distractions from what you should be doing, and not to be dallied with. This negative attitude toward psi and the *siddhis* pervades the Vedantic tradition and has now passed through to Western transpersonal psychology, with the result that many of the people that we and other sympathetic scientists would love to work with—advanced meditators, in particular—seem unwilling to become involved even on a strictly limited basis with something they view as not worthy of close attention. One of the things I hope to accomplish with this chapter is to encourage greater openness, in persons who are real adepts in meditation, to the idea that they will not be harmed by participation in research that could go a long way toward empirically validating foundational elements of their traditions.

SOME PERSONAL HISTORY OF PSYCHICAL RESEARCH

Next I would like to say a bit about the history of psi research and how I became involved in it. I will also try to say a bit more about the general state of the field at present, including what has begun to emerge in terms of its theoretical implications, and I will get fairly specific about the things that we are attempting to do in this area at the Division of Perceptual Studies (DOPS) and why we are interested in doing those things. And then I will end with an appeal for collaborative work between serious practitioners of meditation and scientists like ourselves who are sympathetic to their core beliefs.

Question number one, of course, is whether psi phenomena really exist as facts of nature. The answer to that question, in my opinion, is an unequivocal "yes." I personally do not think there is any reasonable doubt about that, although it continues to be a subject of hot debate in some quarters. Like William James, I think the basic factual issue was already settled by the end of the nineteenth century. We have been trying to get beyond that point ever since, not only in terms of more convincingly demonstrating the sheer existence of psi phenomena, but more importantly in trying to understand how they work and what implications they may have for our general scientific worldview.

Let me next quickly sketch the main lines of that history. In its formative stages, following creation of the British Society for Psychical Research (SPR) in 1882, investigators focused mainly on spontaneous cases of telepathy, crisis apparitions, field studies of mediumship, and things of that sort, although there were also significant amounts of experimental work on topics such as telepathy, hypnosis, and psi-conducive psychological automatisms such as crystal-gazing ("scrying") and automatic writing.

A lot of remarkable things got accomplished during that early period, but William James was barely in his grave when radical behaviorism burst upon the scene here in the United States, with unfortunate consequences not only for psychical research but for psychology in general. James had defined psychology as the science of mental life and its relationship to physiological processes, emphasizing real human experience in its endless variety, but his broad vision was essentially pushed aside by the narrow behaviorist alternative advanced in 1913 by John B. Watson.[3] Introspection was to be done away with altogether, and it would no longer be considered helpful or proper to talk about things like consciousness, mental imagery, or anything else having to do with inner experience or the mind. Psychology would henceforth concern itself only with stimuli and responses and their supposedly lawful connections. This radical and in retrospect bizarre doctrine achieved virtual hegemony during the next fifty years of mainstream American experimental psychology. Many have said, and it is probably true to some degree, that psychology was just trying to imitate the methods of classical physics in hopes of achieving something resembling its level of success. Physics, of course, is the paragon of success in the hard sciences, and it is all about finding mathematical equations that describe and predict the behavior of objective things—things that lack experiential interiors and can be studied effectively from an exclusively third-person or "objective" point of view. Hardcore behaviorists apparently hoped that psychology would turn out that way too.

In 1927, with behaviorism already at its zenith, Joseph Banks Rhine (1895–1980) arrived in Durham, North Carolina to take a position in Duke University's brand-new Department of Psychology. He had been invited there by the department's first chairman—the antibehaviorist and antimaterialist William McDougall—specifically to establish a program of experimental research on paranormal abilities. Rhine and his colleagues proceeded to crystallize and validate a family of simple experimental and statistical procedures (mostly based on guessing of hidden geometrical symbols—Zener cards—to study ESP, and dice-throwing to study PK) that became the workhorse tools of experimental parapsychology for decades. They deployed these tools relentlessly, and generated a large volume of peer-reviewed experimental reports that confirmed the reality of the basic psi phenomena and began systematic exploration of factors controlling their expression.

Although Rhine's approach is today sometimes scorned, usually on grounds that it trivializes psi by making it the subject of boring laboratory tests, he and his colleagues accomplished a remarkable amount of solid experimental work. In addition to massive amounts of direct evidence for psi, usually in the form of above-chance scoring on the designated targets, they found a number of other systematic effects that tell equally strongly for the reality of psi. In card-guessing tasks, for example, they repeatedly found (often in data that had originally been collected for other purposes) effects such as "terminal salience" (consistent elevation of scores near the beginnings and ends of runs), "displacement" (consistent scoring not on the designated targets themselves but on specific near-neighbors in the sequence of targets), "psi-missing" (systematic *avoidance* of correct responses), and "consistent missing" (systematic but *incorrect* associations between particular targets and responses). They also effectively countered, often with the help of professional statisticians, all of the methodological issues raised by their early psychological critics. In sum, at the cost of temporarily contracting "psychical research" as broadly conceived by our founders into the narrower discipline now known as "experimental parapsychology," Rhine and his colleagues kept the field alive and moving forward during a very tough period.[4]

Rhine was still going strong in 1970 when I was finishing my PhD, and although I had begun graduate school as a conventional physicalist, passively absorbing the prevailing scientific worldview, I had subsequently experienced two major events that made me more receptive to his work than I would have been earlier. The first was that I had become disillusioned with the so-called computational theory of the mind, which at that time was the big new thing in experimental psychology, rapidly taking over from Watson's radical behaviorism. People had realized that minds are far more complicated than behaviorism in its original form admits, and now we had a technology that allowed us to construct detailed and testable models of what might be going on inside human beings that could account for their behavioral capacities. This was certainly a big improvement, and it appealed strongly to young graduate students like me. But then I started my dissertation research, which had to do with computer "understanding" of natural language, and it quickly became evident that these problems are much deeper and harder than I had been led to believe, and that a purely computational approach was just not going to work.

In addition, and around the same time, one of my female relatives suddenly became a "medium"—i.e., a person who goes into an altered state of some sort and then delivers messages ostensibly coming from "the other side," or from some other department of reality. One day I received a phone call about this from my mother, who clearly hoped I would be able to reassure her. At the time I did not have the slightest idea what she was talking about, and so headed over to the library to find out. There I quickly discovered to my surprise that William James himself had spent a large part of his adult life studying this and related subjects. I had not known that, despite the fact that I was then working in Harvard's own William James Hall, and I doubt that even a handful of the many other denizens of that large building knew it either! In any case, I started to read about mediumship and quickly concluded that my relative was not in grave psychological danger. I also started reading more widely in the experimental parapsychology literature, and began to correspond with J. B. Rhine, who by then had retired from Duke and moved his entire operation off campus. In due course, following a postdoctoral year in computational linguistics with Susumu Kuno at Harvard, I went to work at Rhine's lab in Durham. I had been there scarcely a month when we were visited by a young man named Bill Delmore (BD), who turned out to be one of the best subjects ever to set foot in a parapsychology laboratory. Bill was then a first-year law student at Yale, and he had found his way to one of my former teachers, Irvin Child, who had himself become interested in parapsychology and who knew that I had started working at the Rhine lab. So BD was referred down to us, and became my housemate for the next six weeks—one of the most life-changing experiences I have ever had.

STUDIES WITH AN EXCEPTIONAL SUBJECT

To give you a concrete idea of the sort of things BD could do, I will describe what happened on the very first day of his initial visit. Rhine's Director of Research at the time, Helmut Schmidt, was a physicist who had recently come to the lab from Boeing.

Schmidt had developed a sophisticated electronic psi-testing machine, a technologically improved version of the standard ESP card-guessing protocols. Outwardly it was a sealed metal box displaying just four large distinctively colored lights and four correspondingly colored buttons, plus counters for trials and hits. The subject's trial-by-trial task was to push the button corresponding to the light that would next be chosen by the machine as a target. The machine determined that target, *after* the response had been registered, using detection of a particle released by a radioactive source inside the box to interrupt an electronic switch that was stepping through the four possible choices millions of times per second. Helmut's device thus utilized the most random process we know about in nature in selecting its targets, and in the absence of a subject it routinely satisfied all relevant statistical tests. Most persons working with it would also score right around 25 percent hits, as expected, but

Helmut after a protracted search had found and reported on a couple of people who could consistently do slightly but significantly better—around 27 percent or so over thousands of trials.

With that as background, BD came into our library on the first day of his visit and sat down to introduce himself to J. B. Rhine, his wife Louisa, Helmut, and the rest of us who were working there at the time. Over the next hour or so he proceeded to describe his very unusual life story, and every so often he would reach out and press a button. By the end of our conversation he had made 508 such responses and scored exactly 180 hits. That may not sound impressive, but it is over 35 percent instead of the 25 percent expected by chance, and if you do the math it turns out that the probability of obtaining a result that good or better by chance is about one in ten million. Given that the expected waiting time for an event of probability p is $1/p$, we could well have sat there for something on the order of 10 million hours waiting for such a performance to happen by chance alone!

Now, it is one thing to read a bunch of experimental literature and in that way become almost but not quite convinced that psi is real, and another to witness it up close and personal. And this was just the beginning. We proceeded to work with BD experimentally over the next year and a half, and found that he could do practically anything we asked him to do. We published a series of peer-reviewed papers about this work, which employed not only the Schmidt machine, but a number of other electromechanical devices and a variety of ESP and PK tests using ordinary playing cards. In one long ESP series, for example, in which his task was to identify randomly selected playing cards individually enclosed in opaque black folders, he got *exact* hits at three times the expected rate ($1/52$) and a large excess of *number* hits as well ($1/13$), results that were extremely significant, statistically.

We were interested, of course, not just in how well BD did but in how he did it, and to this end we carried out a parallel series in which we showed him pictures of the same playing-card targets under bad viewing conditions (brief exposures, using a shutter tachistoscope), so that we could study his systematic errors in *visual* identification. As expected, he tended under these conditions to mix up face cards, cards with similar numbers of pips, and cards of the same color (clubs with spades, for example, other things being equal). What was astonishing and gratifying, however, was that in the *clairvoyance* condition, in which he was strictly deprived of visual access to the cards, he displayed extremely similar patterns of systematic error. We interpreted this as confirming a central feature of his self-description: Specifically, to do the task he would close his eyes and wait for involuntary visual imagery to appear; the images would typically be small and fast-moving, poorly lit and colored, and unpredictable as to their point of origin and trajectory within his visual field, and so even when the images themselves correctly encoded the targets he would naturally make secondary errors of identification that resembled his visual errors. In effect, we had succeeded in exposing part of the internal basis of his very unusual ESP ability.[5]

In another study BD's task was to shuffle a "response" deck of playing cards with the intention of matching as well as possible a previously randomized "target" deck. Under these conditions he scored even better in terms of exact hits (four times the rate expected by chance), but the excess of number hits and the pattern of systematic visual-like errors observed in the clairvoyance series disappeared. Most significantly, it appeared that he had to be influencing the order of the cards to some degree, a PK effect, rather than just repeatedly shuffling and clairvoyantly evaluating outcomes produced by chance alone. The argument for this was that he typically shuffled his deck just a few times, whereas he would have had to shuffle it dozens or even hundreds of times in order to obtain by chance the numbers of hits he sometimes got—as many as five, six, seven, or more per run.[6]

TOWARD A NATURAL HISTORY AND PSYCHOPHYSIOLOGY OF PSI

It seemed likely that something special might be going on in BD's brain just before he made a successful ESP response, and this is what started us down the path toward psychophysiological studies of psi. The basic idea was to try to distinguish correct responses from incorrect ones in terms of accompanying EEG (electroencephalograph) signals, which provide measures of the electrical activity of the brain.

We recognized early on that there were a number of ways in which information of this sort would be important and helpful, if we could obtain it. For one thing, it would anchor free-floating psi events to events of another measurable kind, which is always desirable. Given the increasingly biological cast of modern cognitive psychology, moreover, anchoring to brain events is especially desirable. Furthermore, that anchoring in itself might provide a helpful degree of *statistical* control of psi, because in principle we could then go through the EEG recordings of a new experimental series and pick out trials that have the right physiological "look," with the expectation that these preselected trials would display a higher hit rate than the series as a whole. We might even achieve *experimental* control, if those psi-conducive physiological conditions should turn out to be ones that we can stabilize or induce, perhaps by some sort of biofeedback regime. We could also conceivably begin to trace flows of psi-related information through the brain, distinguish between candidate sources of observed psi effects (most notoriously, between the persons nominally identified as "subjects" and "experimenters"), and attempt to discriminate experimentally between competing theories of psi.

We began developing a program of this sort when I moved from the Rhine lab to the Electrical Engineering Department at Duke in mid-1973. One of the important things that we became increasingly aware of during the following years is what might be called the "natural history" of psi, by which I mean in particular the fact that its historical and cross-cultural expressions in human affairs have been anything but randomly distributed in space and time. Unusual outcroppings of strong psi phenomena appear instead to occur conspicuously in conjunction with a relatively few special individuals, often under circumstances involving altered states of consciousness (ASC) of certain definite and related kinds such as deep meditation, deep hypnosis, mediumistic trance, out-of-body experiences (OBEs) and near-death experiences (NDEs), and mystical and psychedelic experiences. On that basis, we advocated for exploration of an indirect psi-research strategy focused initially on the relevant altered states, with the expectation that if we can find people who are adept at entering such states—or better yet, learn how to produce the states ourselves—then psi will come along for the ride. This, we argued, could provide the ultimate solution to the problem of repeatability in psi research.[7]

Striking confirmation of the importance of this hypothesized connection between altered states and psi soon came from a quite unexpected direction, when in 1982 I carried out a massive data-analysis project that had interested me for a long time. Specifically, over the previous ten years I had collected a sizeable number of historically significant high-scoring forced-choice datasets deriving from six different gifted subjects and a variety of psi-testing procedures, and had begun to suspect, mainly through the work with BD, that for him at least and possibly for some of these other exceptional individuals as well, there was a tendency for hits to come consecutively, in long strings or batches, far more than one would expect by chance. As it happens, there is a nice clean statistical way of separating these two questions: (1) whether the overall number of hits is itself significant; and (2) whether whatever hits do occur tend to occur in groups rather than being randomly distributed. I applied these methods exhaustively to my high-scoring datasets, and the answer was dramatically clear: In almost every case the hits tended strongly to come in groups, so much so that in several cases a few scattered long strings of hits clearly accounted for practically all the psi there was in the entire lengthy performance! Moreover, there was striking preliminary evidence of association between these occasional outbursts of unusually strong psi hitting and the occurrence of well-marked shifts in subjects' internal state. For example, some studies included rudimentary physiological measures such as the galvanic skin response (GSR), which suddenly altered just as the episodes began, and subjects themselves sometimes reported feeling somehow "different" and then proceeded to reel off a long string of hits. In sum, significant traces of the hypothesized linkage between altered states and psi could be found even in the old Rhine-type, forced-choice testing situation, which on the face of it seems about the least likely of all possible environments in which to expect such a connection.[8]

Other related developments were occurring at that time within parapsychology more generally. During the 1970s and 1980s quite a bit of headway was made, largely as a result of experimenters learning to adapt their methods better to the natural history of psi. For example, since the early days of the SPR, as confirmed later by the spontaneous-case studies of Louisa Rhine and others, it had been known that a large proportion of spontaneous psi experiences, especially precognitive experiences, occur in connection with *dreams*. It had also recently been demonstrated that the occurrence of vivid dreams in sleeping humans could be detected using rapid eye movement (REM) and EEG criteria. It therefore became natural to think about experiments in which the dream state could be experimentally probed as a vehicle for psi.

A landmark series of studies investigating ESP and dreaming was carried out at Maimonides Medical Center in Brooklyn, New York, which produced statistically strong results in a small number of trials.[9] While a subject was sleeping, an experimenter or "agent" in a separate room focused on randomly selected and visually rich target material such as art prints, with the intention of communicating that material to the sleeping person. When the physiological markers of dreaming appeared, the subject was awakened by a second experimenter blind to the target and asked to report any remembered dream content. Following completion of data collection, the subjects themselves and/or outside judges could then blindly rate these reports for correspondence with targets and decoys, providing a basis for statistical evaluation. Because their results were so striking, these dream-telepathy studies were attacked at the time by a number of disgruntled mainstream psychologists, but their published criticisms were systematically demolished by Irv Child in the pages of a prominent professional journal, the *American Psychologist*.[10]

These results were very encouraging, but of course dream studies themselves are hard to carry out in that they are labor-intensive and require a lot of late-night work. Could we perhaps find a way of getting most of their experimental benefits in a logistically more convenient way? This led to development of the so-called "ganzfeld" protocol, which attempts to get people into an imagery-rich hypnagogic or "twilight-zone" state at the border between sleep and wakefulness.[11] It is worth mentioning here that the authors of *Phantasms of the Living*[12] had pointed out in 1886 that a surprisingly large number of their crisis-apparition cases occurred in such states, far more than would be expected from the proportion of time we humans typically spend in them. The ganzfeld procedure typically involves putting halved ping-pong balls over subjects' eyes and earphones over their ears, exposing them to homogeneous or random light and sound, and taking them through some sort of progressive relaxation procedure. The targets are again rich free-response materials of one or another sort, responses consist of verbal descriptions or drawings, and outcomes are evaluated statistically using blind judging procedures. Many such studies have been carried out, by many different investigators representing a number of different laboratories, and like the dream studies they have generally proved far more efficient than Rhine-style forced-choice card-guessing studies in the sense that they typically produce stronger effects in smaller numbers of trials. In his excellent 2006 book *Entangled Minds*, Dean Radin summarized a number of the main lines of experimental psi research including the ganzfeld, and at that point no less than eighty-eight ganzfeld studies had been published, spanning a period of thirty years, with a cumulative statistical p-value that was vanishingly small.[13]

All this certainly represents real progress in the experimental study of psi phenomena. At the same time, I hasten to add that even the ganzfeld procedures are rather labor intensive. Each session may take an hour or more to complete, consists of just one trial, and typically results, after the judging, in a single four-choice test result. It has also recently been demonstrated using EEG methods that current ganzfeld procedures do *not* automatically or routinely propel subjects into the targeted sleep-border state of consciousness. And to put things in a little different perspective, let me also mention that the gifted subject BD produced, in one week of formal testing at the Rhine lab with the four-choice Schmidt machine, results that are statistically more or less equivalent to that entire thirty-year history of ganzfeld research.[14]

The net result of all these developments was to make it gradually very clear to us that we should focus our own future psycho-physiological research efforts on intensive longitudinal studies with carefully selected participants. We saw clearly that psi is not democratic, that persons would not come to us equally endowed for strong psi performance, and that there was no reason to expect that all gifted subjects would succeed at the same tasks and in the same ways. We also recognized that human brains are far from identical and interchangeable, and that it is necessary to respect individual differences of anatomical and functional sorts as well as psychological sorts. Finally, as described above, we had also become aware of the deep interconnections between psi and altered states of consciousness.[15] We therefore resolved to focus our own future research efforts on persons who manifest unusual skills of various kinds—not just psi-performance skills, but ability to voluntarily enter any of the various kinds of markedly altered states of consciousness that we had identified as psi-conducive: hypnagogic states, mediumistic trance, deep hypnosis, out-of-body experiences, and especially deep meditative states, for which there is a great deal of relevant lore but not much serious scientific evidence—at least not yet.

PROFESSIONAL AND THEORETICAL
INTERLUDE

This was the furthest point of advance of our Duke program as it evolved up through about 1987. We had worked diligently along these lines for over a dozen years and made, I believe, very significant progress, but the computer and EEG technology of that period really was not up to the task, and we had run out of funding. I next took a long and financially mandatory detour through mainstream biomedical research at the University of North Carolina, Chapel Hill, primarily conducting EEG and fMRI (functional magnetic resonance imaging) studies of adaptation to natural tactile stimuli in the human somatosensory cortex. This afforded excellent opportunities to learn about emerging developments in systems neuroscience while also allowing us to continue developing and validating research tools that we intended ultimately to apply to problems in experimental parapsychology.

In 2002 I moved to the University of Virginia in Charlottesville in order to resume working full-time in psychical research. The first order of business was to intensify effort on a project that had begun in 1998 under the auspices of the Center for Theory and Research (CTR) of Esalen Institute in Big Sur, California.[16] Esalen's co-founder, Mike Murphy, had recognized that postmortem survival is a theoretically pivotal issue in the contemporary science/religion debate, because survival beliefs are common to the world's religious traditions but cannot be true if the current mainstream scientific worldview is correct. To address this issue he had therefore assembled an initial working group of about twenty scientists and humanistic scholars involved with the subject professionally, including several of us from DOPS.

We devoted the first two of our annual five-day meetings mainly to presenting and discussing currently available evidence for survival, but by the end of the second meeting a much more ambitious two-stage plan had emerged: We would first attempt to undermine the physicalist metaphysics that undergirds virtually everything now going on in psychology, neuroscience, and philosophy of mind, and then go on to identify at least approximately what sort of alternative conceptual framework or metaphysics will be needed in order to overcome its deficiencies.

Let me begin by explaining more precisely what it is that we all found so unsatisfactory about the current mainstream view. Classical physicalist metaphysics is the modern philosophical descendant of the "materialism" of centuries past. It comes in a variety of subtly different shadings, but the basic story common to all of them goes like this: *All* facts are determined in the end by physical facts alone. Reality consists at bottom of tiny bits of self-existent stuff moving in accordance with mathematical laws under the influence of fields of force, and everything else must derive somehow from that basic underlying stuff. In particular, we human beings are nothing more than extremely complicated biological machines. Consciousness and its contents are generated by (or in some mysterious way identical to, or supervenient on) neurophysiological events and processes in the brain. Our everyday experiences of mental causation, free will, and the self are mere illusions, by-products of the grinding of our neural machinery. And of course since our minds and personalities are entirely products of that machinery, we are necessarily extinguished, finally and totally, by the death and dissolution of our bodies. On a more cosmic scale, there are no final ᵃ⁻
there is no transcendent order: The overall schem
utterly devoid of meaning or purpose.

This bleak worldview has permeated the opinion elites and educational systems of all advanced societies, and is undoubtedly a principal driver of the pervasive "disenchantment" of our modern world with its multifarious and worsening ills. Reasons for doubting its truth, however, are rapidly gaining cumulative force. In the first place, classical physicalism is not merely incomplete, which nobody can deny, but incorrect at its very foundation in physics itself. Among other things, the deterministic clockwork universe postulated by Newton and Laplace was overthrown with the rise of quantum theory a century ago, and "matter" as classically conceived has been shown not to exist. The prevailing physicalist "production" model of brain/mind relations now seems headed in the same direction: At present we have no understanding whatsoever of how consciousness could be generated by physical events in brains, and recent theoretical work in philosophy of mind has convinced many, including at least a few prominent neuroscientists, that we can never achieve one.

In sum, there already seems to be increasing recognition in many quarters that we are at or very near a major inflection point in modern intellectual history. What our Esalen group set out to do in support of this impending revolution was to assemble in one place large amounts of peer-reviewed empirical evidence for a variety of human mental and psychophysical capacities that resist or defy explanation in conventional physicalist terms. Evidence for psi and survival of course comes first on any such list, because of its profound and inescapable conflicts with conventional physicalist expectations. It is for precisely this reason that many mainstream scientists are anxious to dismiss it, or perhaps more accurately to isolate and quarantine it as though it represents the only arena in which contemporary physicalism is not triumphantly advancing. In fact, however, there are many other unexplained empirical phenomena that point in the same direction.

We approached the task of assembling this larger body of relevant evidence by revisiting an extraordinary but neglected book, published in 1903, which had already pursued the same general strategy: *Human Personality and Its Survival of Bodily Death*, by Frederic W. H. Myers (1843–1901), a founder of the SPR and friend and colleague of William James.[17] We set out to update and re-evaluate Myers's great work in light of the subsequent century of peer-reviewed scientific research on various topics that had been central to his own original argument. To this end we systematically collected material related to manifestations of extreme psychophysiological influence, such as stigmata and hypnotically induced blisters; prodigious forms of memory and calculation; unexplained aspects of everyday human memory; psychological automatisms and secondary personalities or centers of consciousness; OBEs and NDEs, including experiences occurring under extreme physiological conditions such as deep general anesthesia and/or cardiac arrest; genius-level creativity; and mystical-type experiences whether spontaneous, pharmacologically induced, or resulting from transformative practices such as intense meditative disciplines of one or another sort. This massive first-stage effort culminated in 2007 with publication of our 832-page book *Irreducible Mind*.[18] Its main conclusion is that an expanded Myers-James "filter" or "transmission" model of human mind and personality, according to which the normal operations of the brain do not *generate* the conscious mind but rather *condition its operation*, not only remains logically viable but has actually been greatly strengthened, empirically, by the additional supporting evidence now available. We were also able to show in surprising detail, I think, how such a picture can be reconciled

;-edge neuroscience and physics, and sketched pos-
ɔns for future theorizing.

I wish to add here that for me personally this first phase
of our project went a long way toward dissolving what the
eminent American psychologist Gardner Murphy had long
ago characterized as the "immovable object" in the survival
debate—the *a priori* biological objection to survival. There can
be no doubt that *if* mind and consciousness are manufactured
entirely by neurophysiological process in the brain, it follows inescapably that postmortem survival is impossible. But
the evidence we assembled in *Irreducible Mind* clearly dem-
onstrates, I believe, that the connections between brain and
mind are in fact much looser, and can be reconceptualized in
the alternative fashion of filter or transmission models without violence to other parts of our scientific understanding of
nature. For me this shift in theoretical perspective immediately
reopened the door to survival as a *logical* possibility, and thus
rendered the existing empirical evidence for survival even more
persuasive than it had previously seemed.

The arguments and data assembled in *Irreducible Mind* not
only invite but *demand* a major overhaul of the prevailing physicalist worldview, and it is essential to recognize here what is at
issue is *not* whether we will have metaphysics—because we inev-
itably will, whether conscious of it or not—but whether we will
have good metaphysics or bad.

Classical physicalism is certainly inadequate, but what sort
of alternative worldview or metaphysics should take its place?
Our basic strategy in approaching this second and much more
difficult task was to examine in depth a number of existing conceptual frameworks or theories, both ancient and modern, that
take the existence of rogue phenomena of the sorts catalogued
in *Irreducible Mind* for granted and attempt to imagine how

reality must be constituted in order that things of those sorts can happen. This effort led to our publication in 2015 of another large book, *Beyond Physicalism*.[19] Our central conclusion there was that theorizing based upon an adequately comprehensive empirical foundation of the sort set forth in *Irreducible Mind* leads inescapably into metaphysical territory partly shared with the world's major religious traditions. Specifically, we argued that emerging developments in science and comparative religion, viewed in relation to centuries of disciplined philosophical theology, point to some form of *evolutionary panentheism* as our current best guess about the metaphysically ultimate nature of things.[20]

In brief, panentheism attempts to split the difference between classical theism and pantheism, conceiving of an ultimate consciousness of some sort as pervading or constituting the manifest world, as in pantheism, but with something left over, as in theism. The version we tentatively embraced further conceives the universe as in some sense slowly waking up to itself through biological evolution in time, as more complex organisms permit fuller expression of the inherent properties of that antecedently existing greater consciousness. Most importantly, the rough first-approximation picture we have developed so far can be elaborated and tested through many kinds of further empirical research, especially research on meditation and psychedelics as pathways into "higher" or "expanded" states of consciousness. In sum, although a great deal remains to be done both theoretically and empirically to bring the current rough picture into sharper focus, we feel confident that it is headed in the right general direction.

What we see currently emerging, in short, is a middle way between the warring fundamentalisms—religious *and* scientific—that have so polarized recent public discourse; specifically, an expanded science-based worldview that can accommodate empirical realities of paranormal and spiritual sorts while also rejecting rationally untenable "overbeliefs" of the sorts targeted by contemporary scientistic critics of the world's institutional religions. This emerging vision is both scientifically justifiable and spiritually satisfying, combining the best aspects of our scientific and religious heritage in an effort to reconcile these two greatest forces in human history. What is ultimately at stake here seems nothing less than recovery, in an intellectually responsible manner, of parts of our human cultural heritage that were prematurely discarded with the meteoric rise of modern science starting four centuries ago. And what is especially significant at this critical juncture, and the fundamental new factor that may finally allow this recovery to succeed after numerous previous failures, is that it is now being energized by leading-edge developments in science itself. Availability of an improved worldview does not guarantee its acceptance, of course, and even widespread acceptance would not guarantee that its potential benefits will be fully realized, or its potential abuses adequately controlled. But a potentially viable path to a better world does seem to me to be opening up.

THE NEW DOPS PSYCHOPHYSIOLOGY LABORATORY AND VISION FOR RESEARCH

Let us descend now from these somewhat dizzying theoretical heights to DOPS's own ongoing research efforts. In 2007, just after publication of *Irreducible Mind*, we were fortunate to receive a sizeable private donation that enabled us to acquire greatly improved space for our entire operation, while also beginning construction of a state-of-the-art neuroimaging facility having capabilities we could only dream about back in our Duke era. In a nutshell, the nascent psychophysiology research program described above has now been reconstituted in greatly strengthened form, with several members of our original research team once again actively involved, and we can at last follow through on the research vision we systematically elaborated during that earlier period.

The lab includes an electromagnetically shielded room and a high-quality commercial EEG system (Biosemi) that allows us to record up to 128 channels of EEG plus other physiological signals such as heart rate, skin conductance, peripheral blood flow, and temperature, plus output from other devices such as PK "detectors" of various sorts, all digitally sampled at high rates and in strict temporal synchrony. In addition, we have software, some commercial and some custom-built by ourselves, which is capable of performing virtually all forms of data reduction and analysis currently deemed helpful for EEG research. We also have access to a magnetic resonance imaging (MRI) scanner dedicated to research at the University of Virginia, and my engineering colleague Ross Dunseath has developed a unique technology that allows for collection of artifact-free EEG concurrently with fMRI data.

With these technical foundations now fully in place, our highest-priority need is to find suitably gifted persons with whom to work, and here I will just quickly sketch a few of the things we intend to pursue in the immediate future. First, we want very much to continue studying persons who can perform strongly and consistently in controlled psi tasks, and to that end we have developed a practice area in the laboratory that is available for people to come in and use, with computer-controlled ESP and PK experiments of various kinds ready to exercise *ad lib*. Although we might find some good psi subjects by directly screening in this way, recent experience with web-based psi-testing has also encouraged us to try a complementary approach. Dean Radin in particular has done a lot of online psi-testing in recent years, involving thousands of people and millions of trials, and it has turned out that one of the most striking characteristics of the few high-scoring people he has found that way is that they also had extensive histories of spontaneous psi experiences. In light of his experience we are now inclined to think that it will probably prove more efficient to work from the spontaneous-experience side rather than expecting more or less randomly selected persons to discover that they have unusual psi abilities simply by trying our tests, whether in the lab or remotely. A rich history of spontaneous psychic experiences is clearly one of the best predictors of capacity for performance in controlled psi tasks, and so we are eager to connect with any of you out there who have such histories and are willing to participate in research. We are also interested, of course, in connecting with persons who have already demonstrated capacities for psi performance under well-controlled experimental conditions, and we would be grateful to hear from or about persons of *that* sort as well.

We also intend to pursue the indirect approach described above, which focuses initially on psi-conducive altered states of consciousness and only secondarily on psi performance. For this reason we are eager to connect with advanced meditators, hypnotic virtuosos, deep-trance mediums, and persons who have had powerful OBEs or NDEs. I emphasize again that we are interested in studying these altered states of consciousness themselves, as well as their hypothesized connections with psi. We would be extremely interested, for example, in persons who can induce OBEs voluntarily, particularly if they can do so from the waking state and then "travel" to a pre-agreed location such as a physically remote "satellite" experimental room coupled to our main lab by optic fiber. Our hope is to find persons who can voluntarily leave their bodies and depart from our shielded room as we continuously record their physiology, "travel" to that satellite room and take note of target materials that have been placed there (and perhaps influence some special detector devices also deployed there), and then come back and report to us what they observed and did while "out." Experiments of this sort could quickly and dramatically expand our knowledge of OBE states themselves, while also adding significantly to the body of evidence supportive of survival.

This brings me back full circle to Ligmincha, Buddhism, and the spiritual traditions more generally. Given the many wonderful things that our new laboratory capabilities in principle enable us to do, our key problem now is to decide which of the available opportunities are really most worth pursuing. The most critical factor by far for us at this point, and the one that will have greatest impact on our trajectory from here forward, is what sorts of unusually skilled people we can bring into the lab for intensive collaborative study. Among all the altered states of consciousness (ASC)/psi connections described above, the one that interests me personally the most, and that I am prepared to bet on as the most fruitful and important of the lot in the long run, is the nexus linking intense practice of meditation in various forms with mystical experiences and psi. Research on meditation has intensified dramatically in recent years thanks to increasing recognition of its potential benefits for human health and well-being, which in turn has provided access to the normal sources of biomedical research funding such as the National Institutes of Health (NIH). Among other things we now know that the brain is far more "plastic" and susceptible to experience-dependent modification than previously thought, and even elementary practice of meditation has already been shown to produce lasting changes in the thickness of various cortical and subcortical structures. More importantly, we already have good evidence that deep meditative states can be physiologically distinctive in surprising ways. Much more surely remains to be learned about meditation itself as research advances along these lines, and that in itself will be a very good thing.[21]

In addition, the world's spiritual traditions are absolutely saturated with lore about the connection between increasing proficiency in meditation and the emergence of *siddhis*, including psi phenomena, and speak about this in substantially common ways. These connections too surely merit deeper investigation, and we want to pursue them in our lab. I can certainly understand to some degree why the psi-research aspects of our program might seem relatively problematic to some spiritually oriented meditators, for Patanjali was certainly correct that excessive preoccupation with *siddhis* can be a manifestation of egoity for persons predisposed that way and hence become an obstacle to self-development. However, I also think it should be possible to practice meditation primarily for spiritual or self-development purposes and yet still devote some time and energy to research on the *siddhis* without thereby distorting one's *sadhana*, especially in view of how much scientific and cultural good could potentially come from such efforts.

At least one significant historical example of such a policy already exists: Specifically, it has recently come to light that the highly regarded modern Tantric mystic and sage Sri Aurobindo (1872–1950) recorded extensive albeit very informal psi "experiments" of his own in his private meditation journals,[22] a preliminary analysis of which has been provided by Richard Hartz.[23] One inescapable conclusion from what has been unearthed so far, however, is that such explorations are more likely to be fruitful if carried out in collaboration with persons well-versed in modern experimental methods. For this reason I wish to appeal to accomplished meditators anywhere who may read these words: *Please consider working with scientific groups such as ours that are genuinely interested in what you do, and that are both ready and able to study it sympathetically!*

Mystical experience and the *siddhis* are deeply interconnected both historically and psychologically, and both are expected consequences of the serious practice of meditation.[24] Not much scientific work has so far been done to explore these connections in the needed depth and breadth, but even though we have barely scratched the surface the results already look promising.[25] Modern psychophysiological methods can teach us a lot about what is going in the minds, brains, and bodies of persons who are adept in meditation, and we can potentially expand that growing body of knowledge in significant ways.

CONCLUSION

Let me end by restating the key points I hope readers will take home from this chapter and from our book as a whole. Psi phenomena definitely exist as facts of nature, and they are accessible to investigation using conventional scientific methods, including methods of the psychophysiological sorts emphasized in our DOPS neuroimaging laboratory. With the conceptual and technical foundations of our program now fully in place, we have every reason to feel confident that we can make significant progress in the experimental directions outlined above.

Psi phenomena are humanly significant in their own right, but they also contribute crucially to a larger-scale scientific and cultural transformation of much greater long-term import. Together with direct evidence of the sorts described in previous chapters for postmortem survival, and in combination with the many other unexplained mental and psychophysical phenomena catalogued in *Irreducible Mind*, they help to demonstrate— *empirically*—that the currently prevailing mainstream scientific worldview needs to be expanded in directions that will bring it into closer relationship with traditional spiritual worldviews including that of Buddhism in its many forms. Because of their potentially decisive significance in forcing this momentous advance, these phenomena deserve to move toward the center of the evolving contemporary dialogue between Buddhism and science.

6

AN EXPANDED CONCEPTION
OF MIND

DAVID E. PRESTI

The contemporary encounter between Buddhism and
science is largely represented—at least in popular
perception—by the study of how meditative practices impact human biology. As they have manifested in the
Western world, these practices are predominantly derived
from Buddhist and Hindu traditions.[1] While the study of
meditation in the context of biophysical science has gone on
for more than fifty years,[2] the attention drawn by the more
recent dialogue between the Dalai Lama and scientists has
elevated this area of research to a place of greater appreciation in the mainstream scientific community. A variety of
projects related to meditation, neuroscience, and psychology
are ongoing—and recent studies include the effects of various
meditation practices on the anatomical structure and neural
activity of the brain, and on attention, memory, sleep, mood,
immune function, and other psychological and physiological
parameters.[3]

In 2012, the first International Symposium for Contemplative Studies brought together researchers from all over the
world working in the fields of neuroscience and psychology

of meditation.⁴ In one of the discussions at this conference, a Buddhist monk and a neuroscientist had a notable exchange.⁵ The monk raised the point that scientific investigation of mind may benefit from paying attention to what he called the "weird things." He then made reference to studies of cases of the reincarnation type (CORT) and to near-death experiences (NDEs). He specifically mentioned Ian Stevenson. He also mentioned telepathy, providing a personal example of an event that took place with one of his teachers.⁶ The neuroscientist stated that if these weird things are actually happening, then "neuroscience is really in deep trouble." Why deep trouble? Because, he said, we would have no way to explain such phenomena.

A CHALLENGE TO EXPAND METHOD AND METAPHYSICS

In this book we describe a variety of phenomena not readily understandable within the current explanatory framework of biophysical science. These are some of the weird things. Such things are also sometimes called anomalous, or rogue, because they are irregular; they do not fit within the current understanding of what orthodox scientific theory holds reality to be. William James once referred to them as "wild beasts of the philosophical desert."⁷ Our contention is that such inconsistencies between observed phenomena and known explanatory mechanisms are actually reasons for excitement—likely to presage new advances in the scientific investigation of mind and its relation to the world of matter.

William James, with customary elegance, put it this way: "Wh̶ was not the science of the future stirred to its conquering activities by the rebellious little exceptions to the science of the present?"[8] And further:

> Round about the accredited and orderly facts of every science there ever floats a sort of dust-cloud of exceptional observations, of occurrences minute and irregular and seldom met with, which it always proves more easy to ignore than to attend to. . . . Anyone will renovate his science who will steadily look after the irregular phenomena. And when the science is renewed, its new formulas often have more of the voice of the exceptions in them than of what we supposed to be the rules.[9]

Many of the phenomena described in the preceding chapters do require an expansion or revision of the explanatory description of nature provided by contemporary physical theory. In such a proposed expansion, everything we currently know from experiments and observations in neuroscience remains valid. What is in "deep trouble" is the *assumption* (or, in more humble terms, hypothesis) that what we call mind or consciousness is *completely* explicable in some straightforward and localized way by neural processes in the brain and body.

Since the phenomena described in this book are considered anomalous, evidence supporting their existence ought to be as convincing as possible. This generally comes only when many researchers are working on problems, when alternate approaches of investigation are employed, when there are independent replications of controversial results, and replications of replications. Even so, no matter how good the experimental results might be, there are more than a few in the scientific community who would still refuse to acknowledge them as meaningful. The eminent nineteenth-century physicist and physiologist Hermann von Helmholtz (1821–1894) was quoted by his colleague physicist William Barrett (1844–1925), both of them Fellows of the Royal Society, in the context of a discussion about telepathy:

> I cannot believe it. Neither the testimony of all the Fellows of the Royal Society, nor even the evidence of my own senses, would lead me to believe in the transmission of thought from one person to another independently of the recognized channels of sensation. It is clearly impossible.[10]

What can be said to that? Perhaps opinions such as these will fall away as more research is accomplished. However, it is difficult at present to accomplish such research, for a variety of reasons. Funding is certainly one. Institutions such as the National Institutes of Health (NIH) and the National Science Foundation (NSF) that fund most university-based biomedical research in the United States will generally not fund these kinds of projects, which at present are funded almost entirely (and somewhat erratically) by private donations.

In addition, and importantly, relatively few scientists are currently working on these questions. Indeed, for scientists to even express interest in these topics can be damaging to their professional reputations. And while students are intrigued when they learn about the study of these phenomena, they are generally not attracted to working on projects, as it is extremely difficult to develop a professional career in this direction.

Often, no matter how carefully done an investigation might be, it is difficult to publish research on these topics in widely circulated, top-tier academic journals.[11] Publication is for the most part in specialty journals having a more restricted readership and far lesser prestige in academia.

Occasionally, however, it does happen that work on psi phenomena is published in major academic journals. Two examples are from prominent social psychologist Daryl Bem of Cornell University. In 1994, Bem published an article—in the journal *Psychological Bulletin*—presenting a comprehensive review of the use of the ganzfeld procedure in the investigation of anomalous information transfer.[12] And in 2011, in the *Journal of Personality and Social Psychology*, he published a series of experimental studies of precognition and premonition.[13] The latter article is especially notable, in that it riled individuals in the mainstream scientific community even *before* it was published.

Shortly before the 2011 article appeared in print, the *New York Times* featured a front-page story on the subject entitled: "Journal's paper on ESP expected to prompt outrage."[14] One academic psychologist was quoted in the *New York Times* article as saying: "It's craziness, pure craziness. I can't believe a major journal is allowing this work in. . . . I think it's an embarrassment for the entire field." And, in a related *New York Times* piece, another prominent academic cognitive scientist stated: "If any of his claims were true, then all the bases underlying contemporary science would be toppled. . . . implications would necessarily send all of science as we know it crashing to the ground."[15]

Certainly it is good that Bem's research received wide attention and later prompted projects to replicate and extend the work, still very much underway.[16] This is how experimental science ought to work. Who knows what is going on with these indications of precognition? The explanation of the phenomena may turn out to be profound or more conventional, but only further research is likely to resolve any questions.

I make no claims as to the veridicality, reproducibility, or interpretation of Bem's data, but cite this as an example of reaction to the mere proposal that something like precognition may be an actual phenomenon, amenable to scientific investigation. Really—is "science as we know it" so fragile and threatened that it risks being sent "crashing to the ground"? It is clear that investigations of topics such as those discussed in this book can evoke very strong emotional reactions. No wonder that even mere mentions of these phenomena—let alone serious discussion—are absent from nearly all contemporary books on psychology, neuroscience, and philosophy of mind. Research into these anomalous phenomena currently lacks a critical mass of scientific community.

Contrast this situation with the financial and social support given to research in certain other areas of basic science. Consider the Large Hadron Collider (LHC), currently the world's largest particle accelerator, which opened on the Swiss–French border near Geneva in 2008. It is estimated that the LHC cost nearly $5 billion to construct and has an operating budget of approximately $1 billion per year.[17] The experiments that resulted in claims of discovery of the Higgs boson—a result that has significant implications for elementary particle theory—are estimated to have cost several billions of dollars. When two papers describing the Higgs boson research findings were published in 2012 by two teams of investigators, these papers had nearly three thousand authors *apiece*.[18] What a testament to scientific adventure and collaboration.

The future may see expansion of scientific research on the topics described in this book, but this is not so straightforward. Laboratory experiments with human beings can be more complicated, in their own way, than the kind of work that goes on in connection with the LHC. Experiments in physical science are constructed so as to minimize any impact human involvement may have. Although quantum mechanics tells us that the human-constructed experimental arrangement may have substantial impact on determining the reality of wl sured, there is still a notion that the role of the observer ᴄa.. minimized. In particular, physicists and biologists do not worry about whether the mental state of the experimenter is perturbing their experimental measurement in uncontrolled ways. They believe they can maintain a kind of external and objective view of nature—a sort of "god's-eye view" or "view from nowhere"[19]— from which it is presumed they are measuring aspects of physical reality existing independently of the mind.

However, if mind and consciousness are in *any* way more than simply the brain-dependent subjective experience of an individual—that is, in any way drawing upon not-yet-understood extended interactions—then it is reasonable that mind may have unexpected causal effects on aspects of the experimental process. This makes so-called "experimenter effects" potentially of profound significance and introduces substantial additional complexity to any investigation.[20] Nonetheless, many laboratory studies of precognition, clairvoyance, telepathy, and psychokinesis have been conducted and the results published in peer-reviewed journals.[21] Some, like the psychophysical studies with Bill Delmore (BD) described in chapter 5, allow for the formulation of additional hypotheses that can be further investigated. Still, even the best of these laboratory results pales in comparison with the spontaneous experiences reported in the real (non-laboratory) world, often in the context of emotionally impactful situations.

Even in the study of phenomena well-accepted in mainstream science, there has been recent discussion related to widespread use of questionable research practices and difficulties with replication.[22] Measuring the energies of transition of electrons between orbitals in a molecule is one kind of thing and may be eminently replicable under controlled laboratory conditions, but measuring the behaviors of complex biological systems and specifically of parameters of human performance under controlled conditions appears to be quite another.

Moreover, there are many paths to knowledge acquisition in addition to those provided by controlled laboratory studies. Philosopher of science Paul Feyerabend (1924–1994) put it this way:

> . . . the world which we want to explore is a largely unknown entity. We must, therefore, keep our options open and we must not restrict ourselves in advance. Epistemological prescriptions may look splendid when compared with other epistemological prescriptions, or with general principles—but who can guarantee that they are the best way to discover, not just a few isolated 'facts', but also some deep lying secrets of nature?[23]

One way forward is to pay careful attention to spontaneous phenomena—documenting, verifying, and following up in any ways that contribute to rigorous investigation of the phenomena of interest. For example, as a result of such investigation, NDEs, once considered too weird to be real, are now generally recognized in the medical community. Anyone who has read the relevant literature knows NDEs occur and that they frequently have profound impact on the experiencer's life. Whether they are explainable in terms of known neurobiology, or yet unknown but nonetheless conventional neurobiology—or will require a substantial shift in our explanatory worldview—is what remains the outstanding question. As more clinicians and researchers become familiar with what has been described regarding NDEs, it is likely that future research will continue to expand. Figuring out ways to conduct studies of out-of-body experiences (OBEs), both as they occur in association with NDEs, as well as in other circumstances, seems especially interesting and important.[24]

As described in chapter 3, it is the case that some children talk as if they are relating details of other lives. This phenomenon occurs all over the world and in many different cultures, yet only a handful of researchers are investigating and attempting to determine what is happening in such occurrences. Certainly more could be done, although this is difficult and time-intensive work, and another phenomenon that does not lend itself to controlled study.[25]

Importantly, our hypotheses may be—indeed, are likely to be—simplistic or limiting in significant ways. While such hypotheses have utility in suggesting and guiding research, because this territory is so unfamiliar and difficult it behooves us to be open to radically new ideas. Attempts to conceptualize these phenomena in terms of models like rebirth of a specific personality or information transfer between the dead and the living may be way off the mark. What are suggested by the data are deep interrelationships involving mind, consciousness, and physical reality—connections we simply do not understand within our present scientific explanatory framework.

We argue that the enigmatic nature of these anomalies may point the way toward new scientific possibilities. What could be a more interesting prospect? Recall (from chapter 1) the optimism of Niels Bohr and Erwin Schrödinger back in the 1930s and 1940s, wondering whether life would perhaps prove unexplainable in terms of the physics of the day, and thus might prompt an expansion of the framework of physical and biological science. Max Delbrück was initially attracted into biology (from physics) by the prospect that elucidation of the molecular basis of life might require new laws of nature. He went on to become a founder of what has become the discipline of molecular biology. And what about Richard Feynman's statement about the best places to conduct experimental research being those regions where we do not seem to know what is happening, using an open-minded approach of trying to prove wrong one's cherished conceptions of how things are working?

These folks were all physicists. Is it the case that physicists are more open to weird and mysterious phenomena? Perhaps. Their field has witnessed the only two major revolutions in the last 150 years of Western science: relativity and quantum mechanics. In addition, it is also the case that physicists often investigate aspects of nature that are far removed from ordinary human experience. Certainly that is the case with relativity, quantum mechanics, and elementary particle physics. Establishing the existence or nonexistence of the Higgs boson—while it has big implications for theories of the elementary makeup of matter— is so distant from ordinary human experience that to most of us it seems essentially irrelevant.

In contrast, the issues discussed in this book involve the deepest questions about the nature of who we are—what is mind, consciousness, our very being-ness—and how do we fit in to the rest of what we understand as reality? Thus, the implications of these lines of investigation strike very close to home. We may choose to have faith that our current framework of physical knowledge and mathematical theory, together with the chemistry and biology it spawns, will ultimately account for everything in some straightforward way. However, there are strong indications that things are not so simple. Rather than being exciting, this is not infrequently experienced as deeply disturbing. Disturbing because it draws attention to how uncertain we are, how little we may actually know about what is going on, about the nature of what we conceive as physical reality—the predictable world "out there" that seems so substantial.[26]

There is something comforting in certainty. Religious faith provides this. So does scientific theory. The term "scientism" has sometimes been employed to describe the dogmatic belief that current scientific opinion defines the limits of the possible. Scientism in science is akin to fundamentalism in religion.

The explanatory frameworks constructed by our science give us powerful ways of describing nature, providing understandability and predictability to the world. Nonetheless, it is likely that our perceptions and current scientific understanding are only scratching the surface, the tip of the proverbial iceberg. Our bodies, brains, and nervous systems, together with our perceptual and cognitive capacities (including our capacity to create explanatory theories that organize our observations) have developed over long periods of time and evolutionary tuning, providing us with a particular slice of reality that is conducive to our survival. It is amazing how much we have been able to figure out. But why would anyone seriously argue that what we know now is the whole story, or even anywhere close to the whole story?[27]

BUDDHISM AND AN EXPANDED SCIENCE OF MIND

We humans are curious and inquiring animals. Our ancient ancestors observed patterns in their environment and devised stories to explain their observations. While these practices date back millennia, the contemporary mode of scientific investigation—seeking explanations grounded in mathematical laws of physics applicable to a reified and mind-independent objective reality—is a relatively new endeavor, dating back several hundred years to around the time of Galileo and Descartes. And while addressing the nature of mind with the methods of Western science is very new – going back perhaps 150 years or so – investigation of mind is old, very old. It is likely that humans have been paying close attention to their inner experience—meditating, so to speak—for many thousands of years. And perhaps much more so in the past, when there were fewer distractions, compared with the busy world of today.

Buddhism and other contemplative traditions have been developing refined procedures of introspection and applying them to investigation of mind for more than two thousand years. A great many highly intelligent and dedicated people have been engaged in these investigations. Might not these methods of introspective exploration and their associated philosophical frameworks (worldviews) have led to discoveries and insights about mind and reality that can inform and complement the ideas of Western science? It seems not only possible, but highly likely.

Alan Wallace speaks to the notion that Buddhism has a long history of developing meditation as a precision tool for exploring the mind, suggesting that for millennia "meditators in different contemplative communities across the globe have systematically explored and reported their findings on inner reality and its connections to outer phenomena."[28] Analogous to the telescope as an instrument of profound significance in astronomy, Wallace suggests that the highly refined techniques of meditation described in various ancient Buddhist texts can be conceived as essentially "telescopes of the mind." And just as telescopes have allowed for a wealth of discoveries in astronomy and completely changed our conception of the cosmos, and microscopes have allowed for a multitude of discoveries in biology, massively informing how we conceive of the nature of life, so also the refined states of awareness that may be cultivated in meditative practices permit unprecedented ability to observe and describe the nature of mind.[29]

In the other direction, the tools and analyses of neuroscience have revealed structural and functional properties of sensory perception that do not appear to have been discovered even after centuries of meditative introspection and resulting commentary on sensory perception in Buddhist philosophy.[30] For example, in teaching science to Tibetan Buddhist monks and nuns, we conduct demonstrations that reveal to the monastics their visual blind spots located not far from the center of gaze, as well as their poor color perception in the visual periphery. The blind spot results from the absence of light-sensitive photoreceptor cells in the place where the optic nerve exits the eye. And poor color perception in the periphery results from our having color-sensitive photoreceptor cells concentrated in the center of gaze.[31] The nervous system, however, "fills in" blind spots and poor color perception, so that we are generally unaware of these things. The monastics are always surprised and impressed.

Buddhist analyses of mind and reality based on contemplative inquiry have perhaps encountered some of the same issues that have manifested as intractable puzzles in science: the mind-body problem, the quantum measurement problem, and the problem of how to even discuss the existence of a reality that could be independent of the concepts we bring to the table (our linguistic, cognitive, and cultural frames). Indeed, I suspect these three puzzles are all facets of the same issue: the inextricable enfolding of mind and world, of so-called subjectivity and objectivity. We know the world via the subjective experience of consciousness, and from this knowing we derive the properties of a reality that is assumed to host our own existence—an objective physical world in which we have evolved as conscious biological organisms. How could this *not* lead to

paradox and conundrum? Scholar of Buddhism Robert Sharf
put it this way:

> Mind and world—knower and known—do not denote auton-
> omous domains or frames of reference, or even interdepen-
> dent perspectives. Rather, mind and world logically enfold one
> another; mind is only possible *within* the world, and the world
> only possible *within* the mind. Which is to say that the logical
> relationship among these antinomies is bound in paradox. . . .[32]

Addressing the paradoxes of mind and world within the framework of contemporary science will doubtless benefit from new
ideas, and it is our belief that transdisciplinary scholarship
emerging from the conversation between Buddhism and science
will contribute to this.[33]

Eleanor Rosch, quoted in chapter 1, stated that "what the
meditation traditions have to offer science is not just more data
to plug into the old ways of looking at brains, but a whole new
way of looking"—and then goes on to indicate several ways
in which traditions such as Buddhism challenge the views of
conventional physicalism. These include concepts from Asian
medicine that view the body as interlocking functional systems;
the concept of the yogic or subtle body and associated subtle
"energy" (called *lung* (Tibetan), *prana* (Sanskrit), *qi* (Chinese))
and energy centers, or *chakras*; direct mind transmission from
teacher to student (a kind of telepathy); and death *samadhi*
(Sanskrit, a deep meditative state).[34]

Sometimes, when an accomplished meditation practitioner
is ready to die, they appear to choose when and where this will
happen. They may sit upright, in meditation posture, and die in
that position, during deep meditation.[35] And sometimes, they
appear to not die all at once, but rather pass through a series

of stages that have been observed and described, a process that has been called *tukdam* (Tibetan *thugs dam*) or *mahasamadhi* (Sanskrit). Here is Rosch again, speaking to death samadhi:

> The breath stops; the brain stops, and if the lama is in a modern hospital, the body will be pronounced clinically dead. However, the heart center remains warm, and the body does not behave like an ordinary corpse. Physically, the lama will remain in meditation posture for hours, days, or weeks, the body decaying somewhat around the outside. The important thing is that during that time, the lama continues to teach—and very powerfully. It is as though removal of all the body's living functions, including the brain, has removed a veil or blockage between the teacher and his students, and now his wisdom mind can reach them much better. One of the most remarkable experiences of my life was seeing a teacher in such a state. Note that the lama is still embodied. But it is a different meaning of body than anything we presently imagine.[36]

Much has been written on death and dying from the perspective of Tibetan Buddhism.[37] Among the more widely known writings in the West is *The Tibetan Book of the Dead*—a text (or set of texts) that became popular after a translation into English was first published in 1927, and later republished with prefatory commentary by the Swiss psychiatrist Carl Jung (1875–1961).[38] This book is said to describe the process negotiated when one dies and passes through the *bardo* (Tibetan, in-between or intermediate state) between the death of this life and rebirth into another.[39]

Some consider *The Tibetan Book of the Dead* to be a sort of scientific text, drawing upon rigorous investigations of death and the after-death experience. For example, scholar of Tibetan Buddhism Robert Thurman characterizes the multitudinous

Buddhist texts on this subject as a "scientific literature on death . . . unique in world civilizations."[40] Other scholars, however, do not interpret *The Tibetan Book of the Dead* as an empirical description of an after-death state—but rather as a set of prayers and practices that suggest a structure for the death experience, thereby helping to prepare individuals to die with greater equanimity; and/or a set of funerary practices used by priests to comfort the dying and their family.[41] Perhaps it is all of the above.

The Buddhism–science dialogue is, and ought to be, multifaceted—with participants bringing a diversity of perspectives, tools, and data to the table. In this book we argue that a very important element of the engagement between Buddhism and contemporary science is that Buddhist cosmology takes seriously the possibility of rebirth, of thought transmission (telepathy) between individuals, and of various other ways in which aspects of mind may transcend any current scientifically understood connection with the body. In addition, Buddhist philosophy has explored how consciousness and the world enfold and co-create each other. This makes for a rich landscape of possibilities for examination.

MOVING INTO THE FUTURE

Some say it is risky to even allow the topics discussed in this book to enter the mainstream of the Buddhism–science conversation—they being so controversial to so many Western scientists that even entertaining discussion is threatening to the viability of the dialogue. But these topics are already present, although largely unmentioned, at least publicly. Nonetheless, there was the discussion referenced at the beginning of this chapter, and there have occurred several other instances of explicit discussion at Mind & Life dialogues with the Dalai Lama.[42] Alan Wallace speaks to the importance of studying these phenomena in many of his writings on Buddhism and science.[43] And anyone coming from the Tibetan Buddhist side of the conversation would be familiar with the lore of the *siddhis* and other "weird things."

Since 2004, I have been honored to be associated with the Science for Monks & Nuns project—within which I teach neurobiology and converse about science with Tibetan Buddhist monks and nuns in India, and more recently in Bhutan.[44] The monastics are always interested in the topics discussed in this book and are puzzled as to why they are generally considered off limits for scientific discussion. The future of the dialogue between representatives of the traditions of Buddhism and science, and the resulting cross-fertilization of ideas, is likely to be strongly influenced by the significant numbers of Tibetan Buddhist monks and nuns who are now studying science as part of their monastic education. This is an impressive step into modernity and will empower the monastic community to connect in greater numbers to this ongoing and evolving conversation, which has thus far largely been driven by the Dalai Lama's great energy, enthusiasm, and influence, together with a relatively small group of scientists from the West.

As the Dalai Lama said in 2002, around the time when the program with which I am teaching was just getting started:

. . . at the beginning [of this dialogue between Buddhists and scientists] there were very few of us from the Buddhist side: at first just myself and our two great translators. Now that we have started modern science studies in the monasteries, we have more and more people involved, and this will continue and spread.[45]

And spread it very much has.

Historically, when the topics in this book have been raised in conversations between Buddhists and scientists, discussion is often dampened by claims that these topics represent fundamental differences between Buddhist belief and scientific practice and thus will need to be excluded from further analysis—"bracketed off" to use the phrase quoted in the prologue of this book.

We argue this should not be the case. These phenomena are profoundly relevant to the nature of mind and the nature of reality. They are amenable to investigation using methods of science, and in some cases significant scientific investigation has been ongoing for more than a century. We believe there to be value in making this work available to more scientists, scholars, and other interested readers, and in more easily accessible ways. Some will hopefully be inspired to learn more through their encounter with the material in this book. Some may be even be inspired to undertake new work in these areas. This is what important but controversial material requires—more and deeper attention.

We do not expect to see general acceptance of this field of research, at least not any time soon, by the mainstream scientific community. This research and what it implies may still be too far from currently accepted explanatory theories for that to happen. In this sense, it is an example of what molecular biologist Gunther Stent (1924–2008) termed "prematurity in scientific discovery." Stent called a discovery "premature" if its implications cannot be connected by a series of simple logical steps to contemporary generally accepted knowledge.[46] There are examples of such events throughout the history of science—experimental findings or explanatory hypotheses that were initially ignored or discredited, and then later embraced by the scientific community.[47] Of course, it is part and parcel of what it means to be revolutionary that such ideas and even data may have to contend with being "premature" in this way. If science is to progress in the usual fashion, theory and even metaphysics will eventually expand to accommodate currently inexplicable findings, if those findings stand up to scrutiny.

In this sense, we believe that cognitive science—the scientific study of the mind—is poised for revolution, that this could be the first true scientific revolution since the advent of quantum mechanics nearly a century ago, and that, as William James speculated in 1892, this revolution might require revision or expansion of the natural-science assumptions (metaphysical framework and physical laws) upon which our current understanding of nature is based.[48] Others have also expressed this idea. Here, for example, is mathematical physicist Roger Penrose:

> It is my opinion that our present picture of physical reality . . . is due for a grand shake up—even greater, perhaps, than that which has already been provided by present-day relativity and quantum mechanics.[49]

And a recent statement by philosopher Thomas Nagel:

> Certainly the mind-body problem is difficult enough that we should be suspicious of attempts to solve it with the concepts and methods developed to account for very different kinds of things [namely, an "external" and "objective" physical world]. Instead, we should expect theoretical progress in this area to require a major conceptual revolution at least as radical as relativity theory, the introduction of electromagnetic fields into physics—or the original scientific revolution itself [the era of Copernicus, Galileo, and Descartes], which, because of its built-in restrictions [to wit, the removal of subjectivity], can't result in a "theory of everything," but must be seen as a stage on the way to a more general form of understanding.[50]

In other words, what is coming is likely to be a real doozy of a revolution.

A SUPER NATURAL (BUT NOT
SUPERNATURAL) CONCEPT OF MIND

Chapter 1 began with an operational definition of mind as the totality of our mental experience: thoughts, feelings, and perceptions, as well as our capacity to be conscious and aware. It was noted that this definition of mind is localized to an individual's personal, subjective experience—and is assumed to result from processes confined to the brain and body, and influenced by information acquired via the known sensory modalities: vision, audition, olfaction, gustation, and somatosensation. However, the material discussed in this book suggests we are at a place where more expanded and nuanced hypotheses about mind may be scientifically justified, hypotheses that incorporate nonlocal and transpersonal aspects—mechanisms by which mental states and conscious experience are influenced by processes that go beyond what is currently describable via the known physics and biology of life.

I like the concept of *super natural*, coined by scholar of religion Jeffrey Kripal.[51] Super natural is two words, and the space in between is essential. In *super natural*, the "natural," refers to nature as it is—subject to observation, analysis, and potential understanding by methods that are scientific—not beyond or above nature (not supernatural, not paranormal). And the "super" honors an expanded view of a shared "mindspace" that opens the door to the reality of any number of anomalous or strange phenomena, the understanding of which is beyond, sometimes perhaps unimaginably beyond, our current capacity to provide explanations.

Such an expanded view might *in part* be described by something like an extension of well-accepted physical phenomena, like electromagnetic fields. We are swimming in electromagnetic energy; the entire universe is filled with it. Yet, we have direct sensory access to only a tiny segment of all that energy—seeing what is termed "visible light" with our visual system, and experiencing through our body the thermal heat of what is termed "infrared." Otherwise, we are completely oblivious to all this energy; only instruments of technology reveal its presence. What would someone living two centuries ago have thought if they had encountered radio, television, and cell phones—technologies that utilize invisible radiowaves and microwaves to transmit and receive information? Inexplicable magic, they would say.

Physicists today speculate upon, and endeavor to measure the existence of, pervasive field/particles called WIMPs—weakly interactive massive particles—hypothetical entities that may underlie the "dark matter" of standard model cosmology.[52] As Eleanor Rosch has recently and somewhat whimsically written:

> What would put these presently marginal studies [that is, the phenomena discussed in this book and expanded upon in many other publications] center stage, of course, would be if physics were to discover something measurable about the mind, apart from the brain, that fit within the ever-expanding domain of what is considered material. We now have particles without mass, dark energy, bosons of various types, and, at least theoretically, vibrating strings of energy that constitute the universe—how about massless *mentons* that operate within a mental energy field? Not impossible; we don't know everything.[53]

That such undiscovered aspects of the world might contribute to a deeper understanding of some of the unexplained phenomena discussed in this book is certainly an idea worth considering. Where might openness to entertaining such new "physical" possibilities lead?

panded view, the brain/body remains the vehicle for
experience of consciousness, but the brain/body is
source of what is experienced. Already in the world
as scientifically described, the brain/body receives signals from
without—photons of visible light impact the eyes, dynamic air
pressure variations the ears, wafting molecules the nose, and so
forth—and these signals from without inform the brain/body in
the generation of our mental experiences of sight, sound, and
smell. In an expanded hypothesis something *profoundly more*
is taking place, whereby the brain/body allows for information
from a novel transpersonal or transcendental domain to enter
one's experience, something beyond the already known sensory
mechanisms.[54]

This hypothesis has a long history, developed in different ways
in different cultures, but with many essential features in common.[55] Within Western scientific and religious philosophy, the
idea goes back at least to ancient Greece and Plato, as well as
Plotinus and the Neoplatonic tradition; and to the Abrahamic
prophets of the Middle East. In the East, the idea is central to the
Upanishads and traditions developing therefrom. The Western
philosophical/metaphysical view was articulated at the end of the
nineteenth century by William James, drawing upon and com-
plemented by work of Frederic Myers (1843–1901), F.C.S. Schiller
(1864–1937), and Henri Bergson (1859–1941). It came more widely
to the public's attention through Aldous Huxley (1894–1963),
writing on his experience with mescaline in *The Doors of Perception*.[56] Central to the justification for holding this view is the fact
that it is not in conflict with any of the findings of contemporary
biophysical science. Moreover, there is much to be gained in the
way of explanatory power for otherwise anomalous phenomena
by allowing for serious scientific consideration and investigation
of transpersonal/transcendental hypotheses of mind.[57]

A metaphor that is often applied to this view—that the brain allows for connection with a "greater" transpersonal/transcendental aspect of reality—is that of a transmitter/receiver/filter scenario, or a "transmission" hypothesis.[58] In this view, the brain allows for a connection to—or reception/transmission/filtering of—something more expansive, nonlocal, perhaps universal. And certain states of consciousness make this connection more likely to occur—opening the filter, so to speak—allowing for deeper glimpses into a larger reality. Spiritual practices throughout history speak to this.

One trajectory of access to such states of consciousness may be contemplative practices, potentially giving rise, as summarized by Patanjali in the *Yoga Sutras*, to the various capacities of the *siddhis*. The *siddhis* described by Patanjali include capacities that fall within the categories of what are currently described as psi phenomena: precognition, telepathy, clairvoyance, and psychokinesis among them. Note that in this context of yoga, the *siddhis* are understood as a normal part of the contemplative path, to be noted in passing as one continues to develop more refined states of consciousness.[59]

Another context of access to transpersonal states of consciousness may be found in cultures embracing shamanism and other forms of indigenous spiritual practice tied closely to relationship with nature—Earth, plants, fungi, animals, and the cycles of the cosmos. Shamans, it is said, have access to other ways of knowing not generally available to ordinary folk— communicating directly with animals, plants, and fungi; and accessing realms of existence outside of our ordinary conception of space and time. Shamans are designated by their communities as healers, and use their skills to benefit their communities.[60] It is common to dismiss shamanic claims as being implausible and based on primitive, uneducated, prescientific notions. However, such societies have been cultivating deep connections with these phenomena far longer than European-American culture has been developing biophysical science (and far longer than any of the existing religious institutions), and it may be that we dismiss such claims only out of our ignorance.

Some shamanic traditions have for millennia employed psychedelic plants and fungi—such as ayahuasca, peyote cactus, and *Psilocybe* mushrooms—to catalyze states of consciousness conducive to accessing transcendental realms of knowledge.[61] While psychedelic substances may facilitate access to transpersonal states of consciousness, little scientific investigation of this has yet been conducted.[62] Now that sanctioned human research with psychedelics is again possible and is re-entering the scientific mainstream,[63] the opportunity to investigate these claims in more controlled ways may be available.

And there are the spontaneous cases—mystical experiences;[64] NDEs; children relating detailed information about other, perhaps previous, lives; apparitions in conjunction with death; mediums who appear to spontaneously access other realms of knowing; and individuals who, for unknown reasons, manifest exceptional psychic abilities. Some of these have been the subjects of this book. And there is much more: ranging from contemporary popular culture to millennia of religious history.[65]

All of this may reflect the inextricably enfolded co-creation of mind and world. The future development of its scientific investigation will require radical new perspective, one positioning mind as a central part of nature: a view of the natural world "that is really and truly a super natural world. . . . one that is deeply material and deeply spiritual at the same time."[66]

CODA

This book is motivated by the notion that that the contemporary encounter between Buddhism and science provides a forum in which to productively explore a more central role of mind and consciousness in our description of nature. Going back to the 1600s and the days of Galileo and Descartes, Western science has separated consciousness off—removed mind from the equations of biophysical science. Mind and consciousness are taken to be something akin to add-ons—emerging from and secondary to physical processes in the brain and body, processes that can be understood to operate independently of mind and consciousness. The defining hallmark of what we call science has been its objectivity—some kind of idealized view from outside.

A closer look reveals that this cannot be the case—mind and consciousness necessarily shape everything we believe we know about reality. And yet for all practical purposes, science operates with an illusion of objectivity—as if mind can be separated off, and its irreducible centrality ignored and even forgotten. This has been a powerfully successful trajectory, and undoubtedly will continue to enjoy varying degrees of success. However, at the same time, we believe limits have been reached—limits that will *only* be transcended by embracing the central role of mind in nature. Within current science, quantum physics already hints that consciousness (experience) is inseparable from our characterization of physical reality. But the coupling between observation and physical reality—the so-called measurement problem in quantum physics—has no agreed-upon metaphysical interpretation or solution. Biophysical science—including quantum physics—seems to be at an impasse with respect to progress into the deepest questions concerning the relation of mind and the world.

What can we do toward opening to the centrality of mind in nature? We do not claim to answer this question; nor do we contend there is one-right-way forward into new terrain. Science is best served by pursuing diverse paths and methodologies. In this book, we essentially propose but one: that the Buddhism–science conversation is an important context in which to discuss the empirical study of phenomena suggesting that aspects of mind transcend the physical body in ways not presently captured by the explanatory framework of biophysical science (grounded, as it is, in a metaphysics of physicalism in which mind is excluded from the beginning). Buddhism takes seriously the centrality of mind in nature—honoring the co-created and interdependent enfolding of mind and the world. This, we believe, is a key aspect—perhaps *the* key aspect—of the beauty, power, and potential of the encounter between Buddhism and contemporary science.

NOTES

EPIGRAPHS

Thomas Nagel (b. 1937): professor of philosophy at New York University. Quoted from Nagel (2017).

Max Planck (1858-1947): one of the architects of quantum physics, and first to put forth, in 1900, the notion of what came to be called energy quanta. Quoted from Planck (1933, 217).

Nāgārjuna (~2nd century CE): founder of the Madhyamaka school of Mahāyāna Buddhist philosophy. Quoted from Nāgārjuna's *Mūlamadhyamakakārikā*, chapter 24, verse 19, translated by J. L. Garfield. See Garfield (1995) and Garfield (2015).

FOREWORD

1. This collaboration has led so far to four separate research studies on the mind's ability to enhance physical well-being, and the National Cancer Institute has awarded M.D. Anderson a large grant to continue the study of Tibetan yoga practices for relief of suffering in cancer patients: Cohen et al. (2004); Milbury et al. (2013); Milbury et al. (2015); Leal et al. (2015); Leal et al. (2018); Chaoul et al. (2018).

2. The various *tsa lung* practices are said to enhance digestion, strengthen the body's limbs, lengthen one's life span, promote feelings of happiness and joy, sharpen the senses, and empower speech, among other benefits. See, for example: Sonam, Gyaltsen, & Gyatso (1974).

PROLOGUE: DEEPENING THE DIALOGUE

1. Kennedy et al. (2005).
2. Seife (2005).
3. Miller (2005).
4. Wallace (2003).
5. Lopez (2008); Lopez (2010) contains a concise summary of the Lopez (2008) book.
6. The Dalai Lama has written two autobiographies (1962, 1998) and has been the subject of several biographies. The story of his life in a beautifully illustrated graphic novel: Meyers, Thurman, & Burbank (2016).
7. The first two formal dialogues (1987 and 1989) between the Dalai Lama and small groups of cognitive scientists and neuroscientists were published as: Hayward & Varela (1992) and Houshmand, Livingston, & Wallace (1999). The story of how these dialogues came to be is described in Houshmand, Livingston, & Wallace (1999, 175–80).
8. The website of the Mind & Life Institute contains a comprehensive list of formal meetings from 1987 to the present, related publications, and for some recent events, video recordings.
9. On the cognitive neuroscience of meditation, see, for example: Lutz, Dunne, & Davidson (2007) and Lutz et al. (2015). On the psychology of emotion, see, for example: Dalai Lama & Ekman (2008). In 2012, the first International Symposium for Contemplative Studies brought together researchers from the world over working in many areas related to the study of meditative practice. Subsequent conferences took place in 2014 and 2016.
10. Dalai Lama (2005).
11. The Dalai Lama began to specifically ask for the introduction of science education programs to the Tibetan Buddhist monastic communities in India in 1998–1999. In 1999, the Dalai Lama asked Achok Rinpoche, then Director of the Library of Tibetan Works and Archives in Dharamsala, to undertake the task of introducing science education into the monastic curriculum. In 2000, the first science workshop was delivered through the Library of Tibetan Works and Archives to a group of fifty monk scholars. And in 2001, the Science for Monks project was created through a partnership between the Library of Tibetan Works and Archives and the Sager Family Foundation. This program

has been skillfully guided by Dr. Bryce Johnson since that time. One of us (David Presti) has been teaching neuroscience to Tibetan monastics with the Science for Monks program since 2004. Another program, Science meets Dharma, created by the Tibet Institute Rikon in Switzerland, began teaching science in Tibetan monasteries in India in 2001. And in 2006, the Emory-Tibet Science Initiative was initiated by Emory University in Georgia and the Library of Tibetan Works and Archives. The Science for Monks program (now expanded to Science for Monks & Nuns) and its larger context is beautifully described in a book of essays and photographs: Sager (2012).

12. Mathison (2011).

13. The Templeton Prize has been awarded annually since 1973. Descriptive information can be found at: www.templetonprize.org.

14. For example: Wangyal (2002, xix–xxii); Cuevas (2003, 28) summarizes a contemporary academic perspective on the relationship between Tibetan Buddhism and Bön: "the developments in Buddhism and Bon were separate but simultaneous processes in the later history [beginning circa 1100 years ago] of Tibetan religious culture. Over the centuries the mixture of indigenous Tibetan beliefs and practices with those of Buddhism and Bon has succeeded in almost completely obscuring any distinctions between them."

15. The Division of Perceptual Studies (DOPS) website at the University of Virginia contains a description of the history and current research activities of DOPS. It also contains a list of many of the publications by Ian Stevenson and current DOPS researchers. Kelly (2013) contains a complete bibliography of Stevenson's publications, and an edited selection of representative works together with commentary by the editor, Emily Kelly.

16. "Psi" is used to refer to a class of phenomena in which information appears to pass between a person and the external world despite no known mechanism by which this could take place. Examples include telepathy (direct knowledge of the contents of another person's mind), clairvoyance (direct knowledge of some spatially remote physical state of affairs), and precognition (direct knowledge of future events). These are all considered subtypes of a more general category called "extrasensory perception" or ESP. The scientific study of these phenomena is called "psychical research," sometimes also called "parapsychology." Another term sometimes used in this context is "paranormal."

Unfortunately, this word often carries the connotation that the phenomena so described are somehow beyond scientific investigation. This
is by no means the case with anything discussed in this book. While
many of the phenomena defy explanation in current scientific terms,
they are not incapable of being studied by the methods of science.

17. Houshmand, Livingston, & Wallace (1999, 69–75). In addition to
drawing from his own experience as a Tibetan Buddhist, the Dalai
Lama was familiar with the work of Ian Stevenson and he and
Stevenson conversed during a visit the Dalai Lama made to the
University of Virginia in 1979.

18. In 1992, at a meeting (Mind & Life IV) entitled "Sleeping, Dreaming,
and Dying," there were discussions on gross versus subtle consciousness,
the process of dying, and near-death experiences: Varela (1997, 161–213).
Specific ideas were even discussed as to how one might begin to scientifically investigate aspects of what is called subtle consciousness. And
in 2002, at Mind & Life X, called "The Nature of Matter, The Nature
of Life," there was a brief discussion on the possibility of scientifically
investigating reincarnation: Luisi & Houshmand (2009, 187–89). The
following year, Mind & Life XI took place in an auditorium before a
large crowd of observers at the Massachusetts Institute of Technology.
The meeting focused on discussions of attention, emotion, and mental
imagery—important topics of investigation in mainstream cognitive neuroscience: Harrington & Zajonc (2006). There was no discussion of topics
that would be in conflict with established areas of study in mainstream
science, and that's pretty much how it's been up to very recently. In 2013,
at Mind & Life XXVI, called "Mind, Brain and Matter," there was a
presentation by Matthieu Ricard in which some of the issues in the present book were mentioned. This time, discussion around the subject gave
rise to a bit more interest than in prior Mind & Life meetings, but still no
one quite knew where to take the discussion, since there was general unfa-
miliarity with the extent of the scientific research on these subjects that has
been done: Ricard, M. "To Look at the Mind with the Mind: Buddhist
Views of Consciousness," in Hasenkamp & White (2017, 151–70).

19. Jinpa (2010, 876). "Causal closure" is a metaphysical principle stating
that physical events can only have physical causes. The use of "enclo-
sure" here is nonstandard and may be a typo in the original essay, as its
author uses "closure" a few sentences later in the same article.

20. Jinpa (2010, 878).
21. Lopez (2010, 893).
22. Cabezón, J. I. "Buddhism and Science: On the Nature of the Dialogue," in Wallace (2003, 35–68, 60–61).
23. The word "spirit" derives from the Latin *spiritus*, meaning breath. It is often taken to mean some sort of animating or vital principle, perhaps associated with life, or even with sentience, but not necessarily limited to these associations. It is taken to be "non-material," in the sense that it is not understood as composed of matter and energy, at least insofar as these qualities are conventionally understood. "Spiritual" is often used to denote dimensions of human experience that transcend the ordinary—things that may be considered "sacred," in that they inspire awe or reverence, and are thus afforded special respect, devotion, or veneration. "Soul," is a word of obscure origin, related to the Old English *sawol*, Old Norse *sala*, and Old High German *seula*. There is overlap with how "spirit" and "soul" are conventionally used, and theologians may debate the fine points of how soul and spirit are similar and different. "Soul" is often used to describe some essential immaterial "part" of a human or animal that can transcend the physicality of the body and may in some way survive the death of the physical body.
24. For example: Gould (1997) and Gould (1999).
25. Harrison (2015).

I. SCIENTIFIC REVOLUTION AND THE MIND-MATTER RELATIONSHIP

1. Kuhn (1996).
2. The vast majority of our mental life at any given moment is typically not conscious, typically out of our awareness. In this sense, perhaps, mind is more fundamental and pervasive than consciousness; but in another sense consciousness is perhaps more fundamental, for it is via consciousness that we know that we know, and via consciousness that we make our inferences about the nature of mind and reality, including inferences concerning the existence of conscious and unconscious mental states, and the existence of physical structures such as brains, bodies, trees, and planets.

3. Of relevance to the Buddhism-science dialogue, scholar of Buddhism Jay Garfield writes:

> Most (but not all) Buddhists share with some (but certainly not all) cognitive scientists and some (but not all) phenomenologists a tendency to think of consciousness as a kind of *thing*, or at least as a discrete *property*. We can then ask how many there are (6, 7, 8 . . .?), whether they are physical or not, what the neural correlates of it are, whether machines can have it, etc . . . These questions may or may not be interesting, and may or may not have answers. But they share a common presupposition insufficiently interrogated in either tradition, viz., that there *is* something called "consciousness," "*jñāna*," "*shes pa*," *etc* . . . about which these questions can be meaningfully asked. (2012, 22)

4. Kuhn (1957).

5. Penrose (2004); Kumar (2008); Rosenblum & Kuttner (2011).

6. Stephen Hawking liked to point out that his date of birth was 300 years to the day of Galileo's death (January 8, 1642). For much of his academic career, Hawking was Lucasian Professor of Mathematics at Cambridge University, the position once occupied by Isaac Newton. And Hawking's day of death was 139 years to the day of Einstein's birth (March 14, 1879). Galileo, Newton, Einstein, Hawking—a noteworthy lineage.

7. Hawking & Mlodinow (2010, 81).

8. Darwin and Wallace developed similar ideas concerning variation and selection as the basis for biological speciation. Although Darwin had formulated the essential concepts beginning in the 1830s, he refrained from publishing his ideas until 1858, following the receipt of a letter from Wallace. Writing from Southeast Asia, Wallace asked for Darwin's opinion regarding similar ideas he had independently developed. Colleagues arranged for a joint publication of Wallace's letter and a brief summary of Darwin's ideas shortly thereafter: Darwin & Wallace (1858). A beautiful collection of essays in celebration of the 150th anniversary of the original publication: Gardiner, Milner, & Morris (2008).

9. Bohr (1933).

10. Max Delbrück's central role in the history of molecular biology is described in a collection of essays honoring his sixtieth birthday: Cairns, Stent, & Watson (1966). See also the superb history of molecular biology by Judson (1979).

11. Schrödinger (1944). Subsequent discussions of the impact of this book on the early history of molecular biology can be found in Stent (1968), Judson (1979), Perutz (1987), and Symonds (1987).

12. The discovery of the molecular structure of deoxyribonucleic acid (DNA) has been told as a compelling and candid autobiographical tale—*The Double Helix*: Watson (1968). See also Judson (1979).

13. While most biologists believe it is possible to *identify* something as either living or nonliving, it is not so easy to arrive at a consensus on precisely what it is that *defines* something as being alive. Is a bacterium alive? All biologists say yes. Is a chromosome alive, or a DNA molecule, or an enzyme or other protein molecule? All biologists say no. Is a virus alive? While most biologists say no, some say yes—or at least maybe. See: Pearson (2008). For a concise summary of thoughts on living versus nonliving: Koshland (2002).

14. Delbrück (1986) describes a perspective on the evolution of human mind in a set of posthumously published lectures based on a course he taught on "evolutionary epistemology" at Caltech during the 1970s. I (David Presti) was a student in this class. Zoologist Konrad Lorenz (1903–1989) in 1941 spoke to the role of biological evolution in shaping our perceptual and cognitive capacities: Lorenz, K. "Kant's Doctrine of the a priori in the Light of Contemporary Biology," in Ruse (2009, 231–47). (This is a translation of a 1941 essay by Lorenz published in *Blätter für Deutsche Philosophie*.)

15. William James was a pioneer in developing a modern science of mind. He was the most famous psychologist of his time and the first professor of psychology at Harvard University. The psychology department building at Harvard is named in his honor: William James Hall.

16. James (1890, 4).

17. A rigorous and comprehensive, yet relatively easy to read and understand introduction to cellular and molecular neuroscience: Presti (2016).

18. A classic articulation of the irreducible nature of subjectivity in the mind-body problem is by Nagel (1974). "Explanatory gap" was coined by Levin (1983). "Hard problem" was coined by Chalmers (1995). See Koch (2004) and Koch (2012) for a neurobiological perspective on the difficulty of deriving mental experience from brain physiology.

19. Philosophy of mind is a richly developed field. A recent exposition: Velmans (2009). A collection of classical and contemporary publications: Chalmers (2002). An excellent, up-to-date, online resource: *Stanford Encyclopedia of Philosophy* (http://plato.stanford.edu/).

20. The notion that mind is an inherent property of life was articulated in the context of contemporary biophysical science by Gregory Bateson (1904–1980) (Bateson [2000]) and Humberto Maturana and Francisco Varela (1946–2001) (Maturana & Varela [1980]). This history is beautifully reviewed in Capra & Luisi (2014, 252–74). This notion can be interpreted as a kind of *panpsychism*, at least regarding living organisms.

21. Various hypotheses have been suggested specifying more precisely what is meant by "sufficiently sophisticated analysis of sensory information." For example, there is "integrated information"—a quantitative measure of the complexity of connectivity within a brain's sensory neural networks: Tononi (2008); Tononi et al. (2016). Another is that the uniquely complex structure of cortical neuropil—with densely packed neurons and glial cells, multitudinous axonal and dendritic fibers and astrocytic processes, chemical and electrical synapses, and electromagnetic field interactions (Presti 2016, 217–19)—results in a capacity for global neurodynamic coherence and the emergence of new electrodynamic states having novel properties and casual efficacy: Freeman (2000); Freeman (2015). Still another idea is an "attention schema" wherein consciousness is described as a neural computational model or simulation of the brain's attentional state: Graziano (2013). This latter approach also specifically places the evolution of consciousness in the context of the evolution of complex social behavior—awareness of personal experience and a sense of self going hand-in-hand with the attribution of awareness and inner experience to others.

22. There are thousands of publications related to the interpretation of reality as revealed by quantum physics. Here's a beginning map: Wheeler & Zurek (1983); Penrose (2004, ch. 29); Penrose (1997); Rosenblum & Kuttner (2011); Stapp (1996); Stapp (2009); Stapp (2011); Fuchs & Schack (2013); Fuchs, Mermin, & Schack (2014); Mermin (2014).

23. Greene (1999); Randall (2005).

24. That additional dimensions beyond the usual three of space and one of time may provide a framework for describing otherwise inexplicable phenomena is an idea going back at least to the novella *Flatland*, published in 1884 by English scholar, clergyman, and educator Edwin Abbott Abbott (1838–1926): Abbott (1884). *Flatland* is a classic tale about the value of getting out of a rut and thinking from a new perspective. Physicists conversant with string theory may not be familiar enough with biology, psychology, and the study of consciousness to consider the possibility that all those extra dimensions, if they are to exist, might possibly have explanatory utility vis-à-vis mind. Psychiatrist and deep thinker about mind and consciousness John Smythies considered the possibility more than a half-century ago: Smythies (1967); Smythies (2012). More recently, mathematical physicist Bernard Carr has explored this idea: Carr, B. "Hyperspatial Models of Mind and Matter," in Kelly, Crabtree, & Marshall (2015, 227–73).

25. Wallace (2000).

26. James (1890, 185)

27. James (1890, 424).

28. Wallace (1999); Wallace (2006); Wallace (2007); Wallace (2009); Wallace (2012).

29. Wallace (2012, 70–71).

30. Rosch (1999).

31. Murphy & Ballou (1960); James (1986); van Dongen, Gerding, & Sneller (2014); Blum (2006).

32. Concluding sentences in: James (1892).

33. Feynman (1965, 127, 148, 158).

2. NEAR-DEATH EXPERIENCES

1. Greyson (1998).

2. Heim (1892); English translation by Noyes & Kletti (1972).

3. Moody (2001).

4. For example: Holden, Greyson, & James (2009); van Lommel (2010); Parnia & Fenwick (2002).

5. Greyson (2003).

6. Owens, Cook, & Stevenson (1990).

7. Holck (1978); Shushan (2009, 37–50).

8. Gabbard & Twemlow (1984).
9. Kellehear, A. "Census of Non-Western Near-Death Experiences to 2005: Observations and Critical Reflections," in Holden, Greyson, & James (2009, 135–58).
10. Roberts & Owen (1988).
11. Greyson (1992).
12. Noyes, R. et al. "Aftereffects of Pleasurable Western Near-Death Experiences," in Holden, Greyson, & James (2009, 41–62).
13. Wachelder et al. (2009).
14. Greyson, B., Kelly, E. W., & Kelly, E. F. "Explanatory Models for Near-Death Experiences," in Holden, Greyson, & James (2009, 213–34).
15. Athappilly, Greyson, & Stevenson (2006).
16. Sutherland, C. "'Trailing clouds of glory': The Near-Death Experiences of Western Children and Teens," in Holden, Greyson, & James (2009, 87–107).
17. Parnia et al. (2001).
18. Osis & Haraldsson (1997).
19. Kelly, E. W., Greyson, B., & Kelly, E. F. "Unusual Experiences Near Death and Related Phenomena," in Kelly et al. (2007, 367–421, 380–81).
20. Chawla et al. (2009); Auyong et al. (2010).
21. Borjigin et al. (2013). See critique by Greyson, Kelly, & Dunseath (2013), and reply to the critique by Borjigin, Wang, & Mashour (2013b).
22. Borjigin, Wang, & Mashour (2013a).
23. Dahaba (2005); Myles & Cairo (2004).
24. Goncharova et al. (2003); Yilmaz et al. (2014).
25. van Lommel (2011).
26. See citations in: Borjigin et al. (2013); also Voss et al. (2014).
27. Greyson, Kelly, & Dunseath (2013).
28. Blanke et al. (2002).
29. Holden, Long, & MacLurg (2006).
30. Devinsky et al. (1989).
31. Greyson et al. (2014).
32. Sabom (1998).
33. Kelly, E. W., Greyson, B., & Kelly, E. F. "Unusual Experiences Near Death and Related Phenomena," in Kelly et al. (2007, 367–421, 392–94).

34. Beauregard et al. (2012).
35. West, L. J. "A General Theory of Hallucinations and Dreams," in West (1962, 275–91).
36. du Prel (1889, xxv–xxvi).
37. Kelly, E. W., Greyson, B., & Kelly, E. F. "Unusual Experiences Near Death and Related Phenomena," in Kelly et al. (2007, 367–421); Parnia & Fenwick (2002).
38. Greyson (1981).
39. Stevenson & Cook (1995).
40. Cook, Greyson, & Stevenson (1998).
41. Sharp (1995).
42. Cook, Greyson, & Stevenson (1998).
43. Ring & Cooper (1999).
44. Sabom (1982).
45. Sartori (2008).
46. Holden, J. M. "Veridical Perception in Near-Death Experiences," in Holden, Greyson, & James (2009, 185–211).
47. van Lommel et al. (2001).
48. Thompson (2015, 299–314).
49. Greyson (2007).
50. Parnia et al. (2014).
51. Kelly (2010a).
52. Sutherland (1995).
53. Greyson (2010).
54. Pliny the Elder (1942, 624–25).
55. Steiger & Steiger (1995, 42–6).
56. Kübler-Ross (1983, 208).
57. van Lommel (2004).
58. Parnia (2013).

3. REPORTS OF PAST-LIFE MEMORIES

1. A review of the University of Virginia CORT research program: Tucker (2007). Another perspective on this research: Shroder (1999).
2. The American Society of Psychical Research was founded in 1885 to promote scientific research into psi phenomena.
3. Stevenson (1960).

158 ∞ 3. REPORTS OF PAST-LIFE MEMORIES

4. Eileen Garrett (1893–1970) cofounded, in 1951, the Parapsychology Foundation, an organization dedicated to supporting the scientific study of psychic phenomena.

5. Stevenson (1974).

6. Stevenson (1975).

7. Stevenson (1977).

8. Stevenson (1980).

9. Stevenson (1983).

10. King (1975).

11. Mills, Haraldsson, & Keil (1994, 217).

12. The Committee for Skeptical Inquiry is a group of scientists, journalists, philosophers, and others who are skeptical of any phenomena that appear to require paranormal explanation—that is, explanation beyond the descriptive capacity of our current framework in biophysical science.

13. Sagan (1996, 302).

14. Tucker (2013).

15. Stevenson (1975, 206–40).

16. Stevenson (1997a).

17. Stevenson (1997b).

18. Stevenson (2001).

19. Tucker & Keil (2013).

20. Dalai Lama (1962, 53).

21. Stevenson (1997a, 300–23).

22. Tucker (2005).

23. Haraldsson (2008).

24. Tucker (2005).

25. Stevenson (1977, 235–80).

26. Stevenson (1997a, 212–26).

27. Stevenson (1990).

28. Stevenson (1977, 15–42).

29. Stevenson & Keil (2005).

30. Stevenson (2000).

31. Stevenson (1974, 109–27).

32. Stevenson (1986).

33. Tucker & Keil (2001).

34. Tucker & Nidiffer (2014).

35. Haraldsson (1995); Haraldsson, Fowler, & Periyannanpillai (2000); Haraldsson (2003).

36. Haraldsson (2003).

37. Leininger, Leininger, & Gross (2009); Tucker (2013, 63–87); Tucker (2016).

38. Tucker (2013, 88–119).

39. Tucker (2000).

40. Sharma & Tucker (2004).

41. Brody (1979).

42. Stevenson & Keil (2000).

43. Schouten & Stevenson (1998).

4. MEDIUMS, APPARITIONS,
AND DEATHBED EXPERIENCES

1. Those involved in this endeavor were by no means on the uneducated fringes of society. In the SPR's early decades, its membership included many who were learned scholars and scientists, a number of whom were Fellows of the Royal Society. Among members of the early SPR were: A. J. Balfour, Gerald Balfour, Sir William Barrett, Henri Bergson, C. D. Broad, Samuel Clemens (Mark Twain), Sir William Crookes, Charles Dodgson (Lewis Carroll), Arthur Conan Doyle, Hans Driesch, Camille Flammarion, W. E. Gladstone, Edmund Gurney, Sir Alister Hardy, William James, C. G. Jung, Sir Oliver Lodge, William McDougall, Gardner Murphy, F. W. H. Myers, H. H. Price, Lord Rayleigh, Charles Richet, John Ruskin, F. C. S. Schiller, Henry Sidgwick, Balfour Stewart, Alfred Lord Tennyson, J. J. Thomson, Alfred Russel Wallace, G. F. Watts, and William Butler Yeats. For an account of the founders and early research of the SPR, see Gauld (1968).

2. Gurney, Myers, & Podmore (1886). A complete digital version of this book is available online at the website of Esalen Institute's Center for Theory and Research.

3. Myers (1903). A complete digital version of the 1903 edition, plus five important contemporary reviews by William James and others, is

available online at the website of Esalen Institute's Center for Theory and Research.

4. For an annotated introductory bibliography to the literature of psychical research—early material as well as contemporary research—see the Appendix in Kelly et al. (2007, 645–55).

5. For an excellent historical review of studies of mediumship, see Gauld (1982). A complete digital version of Gauld's 1982 book is available online at the website of Esalen Institute's Center for Theory and Research. The complex debate between the survival-interpretation and the psi-interpretation is discussed by Gauld (1982) and Braude (2003).

6. The number of these so-called "drop-in" cases is not insignificant. Ian Stevenson's unpublished notebooks document that he had identified about sixty-five such cases published in a variety of, sometimes obscure, places. For references to some of the cases that Stevenson and others have published, see Kelly (2010b).

7. Kelly (2010b).

8. Kelly & Arcangel (2011).

9. "My Grandfather's Clock," written in 1876 by Henry Clay Work: "It was bought on the morn, of the day he was born, and was always his treasure and pride; but it stopped short, never to go again, when the old man died."

10. Hart & Hart (1933).

11. Tyrrell (1953).

12. Barrett (1926).

13. Osis & Haraldsson (1997).

14. Fenwick, Lovelace, & Brayne (2010).

15. Stevenson (1995).

16. For specific references to such cases, see Kelly et al. (2007, 410). For visions such as these during NDEs, see chapter 2 of this book, and Greyson (2010).

17. Nahm et al. (2012).

18. Rush (1812, 257).

19. Haig (2007).

20. James (1900).

21. For references to people who have argued in support of this view, see Kelly et al. (2007, 73).

5. PARANORMAL PHENOMENA, THE *SIDDHIS*, AND AN EMERGING PATH TOWARD RECONCILIATION OF SCIENCE AND SPIRITUALITY

1. Gauld (1982); Braude (2003); Sudduth (2016). Additional literature on survival versus super-psi (also called living-agent psi) may be found in an annotated bibliography on psychical research in *Irreducible Mind*: Kelly et al. (2007, 645–55).

2. Patanjali is believed to have lived at least 1700 years ago. The *Yoga Sutras*, the compilation of which is attributed to him, consists of incisive aphorisms systematically setting forth the physical, mental, and spiritual disciplinary practices that constitute yoga. Many translations of and commentaries on the *Yoga Sutras* have been published. The one quoted here is: Taimni (1961), Book 3, Sutra 38. For a more detailed discussion of Yoga in relation to psychical research, see Chapter 9 ("Patanjali's *Yoga Sutras* and the *Siddhis*") by Edward Kelly and Patanjali scholar Ian Whicher in: Kelly, Crabtree, & Marshall (2015, 315–48).

3. Watson (1913).

4. Pratt et al. (1940).

5. Kelly et al. (1975).

6. Kanthamani & Kelly (1975).

7. Kelly & Locke (1981/2009).

8. Kelly (1982).

9. Ullman, Krippner, & Vaughan (1973); reissued with two new introductions by Montague Ullman and Stanley Krippner: Ullman, Krippner, & Vaughan (2003).

10. Child (1985).

11. "Ganzfeld" is German for "whole field" and is used to denote the experimental creation of conditions in which the subject experiences homogeneous, unpatterned sensory input. Such conditions are thought to make it easier for extrasensory input to enter into awareness. A description and review of the ganzfeld procedure can be found in: Bem & Honorton (1994).

12. Gurney, Myers, & Podmore (1886).

13. Radin (2006). More recent updates on the ganzfeld and many other lines of psi research can found in: Cardeña, Palmer, & Marcusson-Clavertz (2015).

14. Ganzfeld EEG results: Wackermann, Putz, & Allefeld (2008). BD's long formal series on the Schmidt machine: Kelly & Kanthamani (1972).
15. Kelly & Locke (1981/2009).
16. Esalen Institute was founded as an education and retreat center on California's Big Sur coast in 1962. Since its inception, it has been at the forefront of developments in humanistic and transpersonal psychology and their connection with societal issues. An excellent biography of Esalen: Kripal (2007).
17. Myers (1903). A complete digital version of the 1903 edition, plus its five most significant contemporary reviews by William James and others, is available online at the website of Esalen Institute's Center for Theory and Research under "Scholarly Resources." See also Kelly & Alvarado (2005).
18. Kelly et al. (2007).
19. Kelly, Crabtree, & Marshall (2015).
20. Murphy, M. "The Emergence of Evolutionary Panentheism," in Kelly, Crabtree, & Marshall (2015, 553–75).
21. See, for example, Lutz et al. (2004), which finds echoes in Buddhist monks of the remarkable early findings of Das & Gastaut (1955) in yogis. See also Kelly et al. (2007, 557–73).
22. Aurobindo (2001).
23. Hartz, R. "The Normality of the Supernormal: *Siddhis* in Sri Aurobindo's *Record of Yoga*," in Rao (2010, 147–68).
24. Marshall (2005); Kelly, Crabtree, & Marshall (2015).
25. Braud, W. G. "Pantanjali Yoga and Siddhis: Their Relevance to Para-psychological Theory and Research," in Rao, Paranjape, & Dalai (2008, 217–43). See also: Kelly, E. F., & Whicher, I. "Patanjali's *Yoga Sutras* and the *Siddhis*," in Kelly, Crabtree, & Marshall (2015, 315–48); and Roney-Dougal, S. M. "Ariadne's Thread: Meditation and Psi," in Cardeña, Palmer, & Marcusson-Clavertz (2015, 125–38).

6. AN EXPANDED CONCEPTION OF MIND

1. The introduction of meditation practices into contemporary Western culture involved a great many teachers from Asia and America. A few key teachers from the Hindu side: Swami Vivekananda, Paramahansa

Yogananda, Jiddu Krishnamurti, Swami Muktananda, Maharishi Mahesh Yogi, Rajneesh, Baba Hari Das, and B. K. S. Iyengar, all from India; and Ram Dass, from America, and through him his Indian teacher, Neem Karoli Baba. And from the Buddhist side: D. T. Suzuki, from Japan; Chögyam Trungpa, Namkhai Norbu, and the 14th Dalai Lama, Tenzin Gyatso, from Tibet; Thích Nhát Hanh, from Vietnam; and American teachers such as Jack Kornfield and Joseph Goldstein, who founded flourishing meditation centers in the United States based on Southeast Asian Theravada Buddhism. Scholars and writers including Aldous Huxley, Alan Watts, Allen Ginsberg, and Huston Smith were key to spreading the messages of meditation and yoga into the popular culture. And the spiritual nexus provided by the Esalen Institute, founded by Michael Murphy and Richard Price, also played an important role. An excellent history of the introduction of Indian spirituality into Western culture is: Goldberg (2010). On Esalen Institute: Kripal (2007). On Huston Smith: Sawyer (2014).

2. Research on meditation, brain and body function, and physical and psychological health was well underway by the 1960s to the 1970s. See, for example: Anand, Chhina, & Singh (1961); Wallace (1970); Goleman & Schwartz (1976); Kabat-Zinn (1982); Benson et al. (1982). A bibliography of thousands of scientific publications on meditation is maintained as a searchable database by the Institute of Noetic Sciences at: www .noetic.org/meditation-bibliography/

3. Recent examples: Tang et al. (2007); Ludwig & Kabat-Zinn (2008); MacLean et al. (2010); Hölzel et al. (2011); Jacobs et al. (2011); Tang et al. (2012); Kaliman et al. (2014).

4. The first International Symposium for Contemplative Studies took place in Denver, Colorado on April 26–29, 2012. Video recordings of presentations at the 2012 meeting can be found on YouTube, archived through the Mind & Life website.

5. The monk was Matthieu Ricard (of the Tibetan Buddhist tradition) and the neuroscientist was Wolf Singer (who investigates brain activity related to perception and conscious awareness). The discussion took place on Saturday April 28, 2012 and can be viewed on YouTube (search for "ISCS 2012—Matthieu Ricard, Evan Thompson, and Wolf Singer—Keynote Address"). Matthieu Ricard's referenced comments

are in the vicinity of 1:11:00 to 1:13:30, and 1:29:00 to 1:32:30; and Wolf Singer's in the vicinity of 1:27:00 to 1:29:00.

Ricard was born in France and after receiving a doctorate in biological sciences, moved to Asia and became a Tibetan Buddhist monk, studying with Dilgo Khyentse Rinpoche (1910–1991), founder of Shechen Monastery in Kathmandu, Nepal, and, for the last several years of his life, the head of the Nyingma school of Tibetan Buddhism. In his presentation, Ricard talked about some of the work explored in the present book and the importance of discussing these topics in the ongoing dialogue between scientists and Buddhists. See also: prologue, note 18.

6. Matthieu Ricard reprised these examples (termed here "irregular phenomena") in a presentation at Mind & Life XXVI, a dialogue with the Dalai Lama on "Mind, Brain, and Matter" in India in 2013: Ricard, M. "To Look at the Mind with the Mind: Buddhist Views of Consciousness," in Hasenkamp & White (2017, 151–70). And he has elaborated again on these and related phenomena, now termed "puzzling experiences," in a recent conversational-format book with Wolf Singer: Ricard & Singer (2017). Ricard has also collaborated with an astrophysicist in an extended conversation about Buddhism and physical science: Ricard & Thuan (2001).

7. James (1909), Lecture 8: Conclusions, final paragraph. This phrase appears near the beginning of the last paragraph of the last of eight lectures delivered by James in 1908 at Manchester College, Oxford: "My only hope is that they [his just-delivered lectures] may possibly have proved suggestive; and if indeed they have been suggestive of one point of method, I am almost willing to let all other suggestions go. That point is that *it is high time for the basis of discussion in these questions to be broadened and thickened up.* It is for that that I have brought in Fechner and Bergson, and descriptive psychology and religious experiences, and have ventured even to hint at psychical research and other wild beasts of the philosophic desert. . . ." [italics in the original].

8. James, W. "The Final Impressions of a Psychical Researcher," in Murphy & Ballou (1960, 309–25). The quoted line is found on p. 325. The essay was first published by William James in 1909, and reprinted in the posthumous collection of essays, *Memories and Studies* (1911).

James goes on to say: "Hardly, as yet, has the surface of the facts called 'psychic' begun to be scratched for scientific purposes. It is through following these facts, I am persuaded, that the greatest scientific conquests of the coming generation will be achieved. *Kühn ist das Mühen, herrlich der Lohn!*" (from Goethe's *Faust*: "Bold is the venture, glorious the reward!").

9. James, W. "What Psychical Research has Accomplished," in Murphy & Ballou (1960, 25–47). The quoted lines are found on pp. 25–26. The essay was first published by William James in *The Will to Believe and Other Essays in Popular Philosophy* (1897). Impressively, James made these statements several years *before* the great paradigm-shifting revolutions in physics of relativity and quantum mechanics took place.

10. Barrett (1904, 328–29).

11. Radin, D. "Seeing and Not Seeing Eternity," in Kakar & Kripal (2012, 208–38); Cardeña (2015).

12. Bem & Honorton (1994); chapter 5, note 11.

13. Bem (2011).

14. Carey (2011).

15. Hofstadter, D. 2011. "A Cutoff for Craziness: Room for Debate – When Science Goes Psychic." *New York Times* (January 6). Online at: www.nytimes.com/roomfordebate/2011/01/06/the-esp-study-when -science-goes-psychic/a-cutoff-for-craziness

16. Bem et al. (2015).

17. Knapp (2012). Online at: www.forbes.com/sites/alexknapp/2012/07/05 /how-much-does-it-cost-to-find-a-higgs-boson/

18. Cho (2012).

19. Nagel (1986).

20. Mauskopf & McVaugh (1980); Hansen (2001); Palmer, J., & Millar, B. "Experimenter Effects in Parapsychological Research," in Cardeña, Palmer, & Marcusson-Clavertz (2015, 293–300).

21. Dean Radin has been a leading experimentalist in psychical research for many years. His books on the subject summarize a great deal of his experimental work (also published in various journals): Radin (1997); Radin (2006); Radin (2013). See also Cardeña, Palmer, & Marcusson-Clavertz (2015).

22. For example: Open Science Collaboration (2015); Baker (2016); Schooler (2014).

23. Feyerabend (2010, 4).

24. Cognitive neuroscience research indicates that perception of one's own body is malleable, and situations can be experimentally created in which perceptual awareness is disconnected from the body—i.e., one can experience their own body from an out-of-body perspective: Ehrsson (2007); Lenggenhager et al. (2007). In these experiments there is no claim that this is anything other than illusion, generated in an experimentally contrived, virtual-reality setup. Nonetheless, such illusions are sometimes cited as proof that out-of-body experiences (OBEs) reported in conjunction with near-death experiences (NDEs) must necessarily be illusions of a similar nature. Perhaps; and perhaps not. Thus, it is important to investigate, to the extent possible, the veridicality of OBEs. See chapter 2, notes 49 and 50. Charles Tart has described several instances of controlled study of OBEs. See Tart (2009, 199–208) describing his work with "Miss Z." A detailed analysis of OBEs from a personal and cognitive-neuroscientific perspective is given by Thompson (2015, 203–29).

25. Mills, A., & Tucker, J. B. "Reincarnation: Field Studies and Theoretical Issues Today," in Cardeña, Palmer, & Marcusson-Clavertz (2015, 314–26); Edelmann & Bernet (2007).

26. In an excellent analysis, social psychologist Mark Leary (2011) articulates and critiques reasons why psychical research is not more a part of the scientific mainstream: mistaken belief that the field suffers from a lack of rigorous research; absence of an explanatory theory or conceptual frame; association with occult beliefs; and discomfort with uncertainty.

27. Cognitive scientist Donald Hoffman and colleagues argue that biological evolution, in the interest of adaptive fitness, may favor perceptual knowledge that is radically disparate from reality: Mark, Marion, & Hoffman (2010). One statement, in Gefter & Hoffman (2016): "Evolution has shaped us with perceptions that allow us to survive. They guide adaptive behaviors. But part of that involves hiding from us the stuff we don't need to know. And that's pretty much all of reality, whatever reality might be. If you had to spend all that time figuring it out, the tiger would eat you."

28. Wallace (2009, 1).

29. For example: Wallace (2009, 82–86).

30. Garfield (2012).

31. For a concise discussion of the relevant visual neurobiology: Presti (2016, 162–64).

32. Sharf (2018).

33. Scholars of Buddhism have begun to explore this terrain: Wallace (1996); Garfield (2015). Cognitive scientists Francesco Varela, Evan Thompson, and Eleanor Rosch have applied frameworks from Buddhist philosophy to investigate mind and its emergent relation to the brain, body, and culture: Varela, Thompson, & Rosch (1991/2016); Thompson (2007); Thompson (2015).

34. Rosch (1999).

35. Several descriptions of the deaths of Tibetan and Bhutanese lamas have been shared with me by senior monastics who were students of the now deceased lamas. Little has been written, at least in English, regarding observations of this phenomenon. One collection of accounts: Blackman (2005). See also: Sogyal Rinpoche (1992, 266–67); Varela (1997, 163–64); Thompson (2015, 293–99). A most impressive analysis is the unpublished thesis from the University of California, Berkeley by Donagh Coleman (2017): "Resting Between Worlds: The Ontological Blurrings of Tukdam."

36. Rosch (1999).

37. Two excellent books: Sogyal Rinpoche (1992); Ponlop (2006).

38. Evans-Wentz (1960). First published in 1927 by Oxford University Press, and re-issued in second, third, and paperback editions in 1949, 1957, and 1960, respectively. The 1957 and 1960 editions contain prefatory commentary by Carl Jung, an English translation of material originally written for the Swiss edition of *The Tibetan Book of the Dead* published in Zurich in 1938. Subsequent to this translation and publication in English by Evans-Wentz, there have been several additional English translations. For example: Fremantle & Trungpa (1975); Thurman (1998); Coleman, Jinpa, & Dorje (2005).

39. The text that has been called in English *The Tibetan Book of the Dead* is, in Tibetan, *Bardo thos grol*, with *bardo* meaning "the between or intermediate state" and *thos grol* meaning something like "liberation through hearing or understanding." Thus: *The Great Book of Natural Liberation Through Understanding in the Between* (Thurman [1998]) or *The Great Liberation By Hearing in the Intermediate States* (Coleman, Jinpa, & Dorje [2005]).

40. Thurman (1998, 17–18), describing Buddha as promoting a rigorously scientific experiential investigation of mind: "It is called 'science' because it is an organized discipline for seeking knowledge of the mind in an exact manner, with a view to freeing individuals from its [the mind's] negative potentials and enabling them to realize its positive potentials. The Buddha set up lifelong educational and research institutions, which eventually developed into what came to be called monasteries and convents. These institutions, dedicated to higher education, spread widely throughout India and the rest of Asia over the centuries after Buddha's time. . . . The study of death, between, and rebirth processes in particular was conducted by researchers within these Mind Science institutions, the results being contained in a huge, cumulative scientific literature on the subject. This scientific literature on death is unique in world civilizations."

And continuing on this theme of ancient Buddhists as scientists of mind, here is Robert Thurman again speaking to tantric practitioners of old: "The Tantric communities of India in the latter half of the first Common Era millennium (and perhaps even earlier) were something like 'Institutes of Advanced Studies' in relation to the great Buddhist monastic 'Universities.' They were research centers for highly cultivated, successfully graduated experts in various branches of Inner Science. . . . I call them the 'psychonauts' of the tradition, in parallel with our 'astronauts,' the materialist scientist-adventurers of the 'outer space' which we consider the matrix of material reality. Inverse astronauts, the psychonauts voyaged deep into 'inner space,' encountering and conquering angels and demons in the depths of their subconscious minds."—Thurman, R. A. F. "Series Editor's Preface," in Gray (2007, ix–xiii).

41. Cuevas (2003) has written a superb history of *The Tibetan Book of the Dead*, in which he teases out the legend positing that the *Liberation Upon Hearing* texts were composed by the legendary yogi Padmasabhava in the eighth century, concealed as a treasure text (*terma*), revealed by the fourteenth-century mystic *tertön* (treasure revealer) Karma Lingpa, and subsequently conveyed to a seventeenth-century adept, Rikzin Nyima Drakpa, who organized and expanded upon the *bardo* prayers and funerary practices, producing what has become the basis for the twentieth-century translations and interpretations of *The Great Liberation Upon Hearing in the Bardo—The Tibetan Book of the Dead*. See also: Lopez (2011); Nahm (2011).

42. Prologue, Note 18; and Note 6 of the present chapter. The Dalai Lama in his autobiography, *Freedom in Exile*, also speaks to developing projects of scientific inquiry into phenomena such as Tibetan oracles (including the Nechung oracle, consulted by the Tibetan government), *tulkus* (a Tibetan term referring to reincarnate lamas), and Tibetan medicine: Dalai Lama (1998, 230–43).

43. For example: Wallace (2007) and Wallace (2012).

44. Prologue, note 11.

45. Luisi & Houshmand (2009, 6). The "two great translators" are Thupten Jinpa and Alan Wallace.

46. Stent (1972).

47. Hook (2002). A classic example of "prematurity in science" is the theory of continental drift—the idea that the earth's continental land masses move relative to one another and once formed a vast supercontinent. Proposed on several occasions over the last four hundred years, it was developed in a now famous paper by Alfred Wegener in 1912, but was largely discounted for at least another four decades. It was simply too weird.

48. Chapter 1, note 32.

49. Penrose (1989, 371).

50. Nagel (2012, 42).

51. Kripal (2016); Strieber & Kripal (2016).

52. While the standard model of Big Bang cosmology is the most successful theoretical framework in cosmology today, its ontological status, along with its various hypothetical entities and phenomena (e.g., dark matter, WIMPS, dark energy, inflation) is not universally accepted. See, for example: Lerner (2004).

53. Rosch, E. "Introduction to the Revised Edition," in Varela, Thompson, & Rosch (2016, xxxv–lv). Rosch's "mentons" reminded me of the "psychons" hypothesized by John Eccles (1903–1997), Nobel laureate (1963) for his role in elucidating cellular and molecular mechanisms of neuronal signaling: Eccles (1990).

54. The additional dimensions postulated to exist in the string theories of modern physics might potentially provide structure to accommodate properties of such transpersonal/transcendental domains. See chapter 1, notes 23 and 24. And certainly there may be other "physical" explanatory structures, not yet discovered, or even imagined.

55. Grosso, M. "The 'Transmission' Model of Mind and Body: A Brief History," in Kelly, Crabtree, & Marshall (2015, 79–113).

56. Huxley (1954). In *The Doors of Perception* Huxley refers to the transpersonal/transcendental domain as "Mind at Large."

57. Kelly et al. (2007); Kelly, Crabtree, & Marshall (2015); Baruss & Mossbridge (2016).

58. Grosso, M. "The 'Transmission' Model of Mind and Body: A Brief History," in Kelly, Crabtree, & Marshall (2015, 79–113); Kelly, E. F., & Presti, D. E. "A Psychobiological Perspective on 'Transmission' Models," in Kelly, Crabtree, & Marshall (2015, 115–55).

59. Taimni (1961); Radin (2013); chapter 5, note 2.

60. Walsh (2007); Harner (2013); Kelly & Locke (2009, 27–40).

61. Harner (1973).

62. Luke, D. "Drugs and Psi Phenomena," in Cardeña, Palmer, & Marcusson-Clavertz (2015, 149–64); Luke (2017); Shanon (2002).

63. For example: Griffiths et al. (2006); Richards (2015); Presti, D. E. "Altered States of Consciousness: Drug-Induced States," in Schneider & Velmans (2017, 171–86); Pollan (2018). In a related vein, two recent books look at connections between psychedelic spirituality and the growth of interest in Buddhism in contemporary Western society: Badiner & Gray (2015); Osto (2016).

64. By "mystical" I mean evocative of profound feelings of sacredness, unity, and connection. Scholar of mysticism Paul Marshall has put it this way: "Perhaps more so than any other kind of experience, mystical experience invites us to question received assumptions about the nature of reality, the ways in which it can be known, and our relation to it. Mystics can feel as if they've looked behind the veil of appearances and caught sight of the nature of self, world, consciousness, time, and even the meaning of it all. While a traditional branch of philosophy called 'metaphysics' has approached a similar set of concerns through discursive reasoning, mystical experience is said to involve a direct intuition, a special way of knowing or 'gnosis' independent of the senses and rational analysis."— Marshall, P. "Mystical Experiences as Windows on Reality," in Kelly, Crabtree, & Marshall (2015, 42, 39–76).

65. Scholar of religion Jeffrey Kripal has written eloquently on the widespread occurrence of the paranormal in the contemporary world, including its integration into popular culture: Kripal (2010); Kripal (2011); Kripal, J. J. "Thinking Anew About Psychical Experiences," in

Kakar & Kripal (2012, xi–xli). See also the personal accounts of clinical psychologist Elizabeth Lloyd Mayer: Mayer (2007).

The possibility that "supernatural" stories of religious history may have more veridicality than contemporary scholars of religion would generally grant them was addressed more than a century ago by William James in his classic work on religious experience: James (1902). Jeffrey Kripal has given this view a greater voice in contemporary scholarship of religion: Kripal (2014); Kripal (2016). Michael Grosso (2016, 2017) has published two thoughtful books on Saint Joseph of Copertino, a seventeenth-century Franciscan friar who was witnessed by hundreds of people over a thirty-five-year period to levitate, sometimes well above the ground and for extended periods of time. Truly weird stuff.

66. Strieber & Kripal (2016, 20). And again, quoting Jeff Kripal (2016, xli): "Such a super naturalism may not be the final answer either. . . . At this point in space-time, we simply do not know. We should thus read that humble space or gap in the middle of 'super' and 'naturalism' not as a certain final answer, but as a hesitation, an act of humility and an opening."

BIBLIOGRAPHY

Abbott, E. A. 1884. *Flatland: A Romance of Many Dimensions*. London: Seeley.

Anand, B. K., Chhina, G. S., & Singh, B. 1961. "Some Aspects of Electro-encephalographic Studies in Yogis." *Electroencephalography and Clinical Neurophysiology* 13: 452–56.

Athappilly, G. K., Greyson, B., & Stevenson, I. 2006. "Do Prevailing Societal Models Influence Reports of Near-Death Experiences? Comparison of Accounts Reported Before and After 1975." *Journal of Nervous and Mental Disease* 194: 218–22.

Aurobindo, S. 2001. *Record of Yoga* (2 vols.). Pondicherry, India: Sri Aurobindo Ashram.

Auyong, D. B., Klein, S. M., Gan, T. J., Roche, A. M., Olson, D. W., & Habib, A. S. 2010. "Processed Electroencephalogram During Donation After Cardiac Arrest." *Anesthesia & Analgesia* 110: 1428–32.

Badiner, A., & Grey, A. (Editors). 2015. *Zig Zag Zen: Buddhism and Psychedelics* (New Edition). Santa Fe, NM: Synergetic Press. (Original, 2002).

Baker, M. 2015. "Is There a Reproducibility Crisis?" *Nature* 533: 452–54.

Barrett, W. F. 1904. "Address by the President." *Proceedings of the Society for Psychical Research, Part 48* 18: 323–50.

Barrett, W. 1926. *Death-Bed Visions*. London: Methuen.

Baruss, I., & Mossbridge, J. 2016. *Transcendent Mind: Rethinking the Science of Consciousness*. Washington, DC: American Psychological Association.

Bateson, G. 2000. *Steps to an Ecology of Mind*. Chicago: University of Chicago Press. (Original, 1972).

Beauregard, M., St-Pierre, É. L., Rayburn, G., & Demers, P. 2012. "Conscious Mental Activity During a Deep Hypothermic Cardiocirculatory Arrest?" *Resuscitation* 83: e19.

Bem, D. J. 2011. "Feeling the Future: Experimental Evidence for Anomalous Retroactive Influences on Cognition and Affect." *Journal of Personality and Social Psychology* 100: 407–25.

Bem, D. J., & Honorton, C. 1994. "Does Psi Exist? Replicable Evidence for an Anomalous Process of Information Transfer." *Psychological Bulletin* 115: 4–18.

Bem, D., Tressoldi, P., Rabeyron, T. & Duggan, M. 2015. "Feeling the Future: A Meta-Analysis of 90 Experiments on the Anomalous Anticipation of Random Future Events [version 1; referees: 2 approved]." *F1000Research* 4: 1188. doi: 10.12688/f1000research.7177.1

Benson, H., Lehmann, J. W., Malhotra, M. S., Goldman, R. F., Hopkins, J., & Epstein, M. D. 1982. "Body Temperature Changes During the Practice of Tum-mo Yoga." *Nature* 295: 234–36.

Blackman, S. 2005. *Graceful Exits: How Great Beings Die—Death Stories of Hindu, Tibetan Buddhist, and Zen Masters*. Boston: Shambhala.

Blanke, O., Ortigue, S., Landis, T., & Seeck, M. 2002. "Stimulating Illusory Own-Body Perceptions." *Nature* 419: 269–70.

Blum, D. 2006. *Ghost Hunters: William James and the Search for Scientific Proof of Life After Death*. London: Penguin.

Bohr, N. 1933. "Light and Life." *Nature* 131: 421–23, 457–59.

Borjigin, J., Lee, U., Liu, T., Pal, D., Huff, S., Klarr, D., Sloboda, J., Hernandez, J., Wang, M. M., & Mashour, G. A. 2013. "Surge of Neurophysiological Coherence and Connectivity in the Dying Brain." *Proceedings of the National Academy of Sciences USA* 110: 14432–37.

Borjigin, J., Wang, M. M., & Mashour, G. A. 2013a. "Reply to Chawla and Seneff: Near-Death Electrical Brain Activity in Humans and Animals Requires Additional Studies." *Proceedings of the National Academy of Sciences USA* 110: E4124.

——. 2013b. "Reply to Greyson et al.: Experimental Evidence Lays a Foundation for a Rational Understanding of Near-Death Experiences." *Proceedings of the National Academy of Sciences USA* 110: E4406.

Braude, S. E. 2003. *Immortal Remains: The Evidence for Life After Death*. Lanham, MD: Rowman & Littlefield.

Brody, E. B. 1979. "Book Review of *Cases of the Reincarnation Type. Volume II. Ten Cases in Sri Lanka* by Ian Stevenson." *Journal of Nervous and Mental Disease* 167: 769–74.

Cairns, J., Stent, G. S., & Watson, J. D. (Editors). 1966. *Phage and the Origins of Molecular Biology.* Plainview, NY: Cold Spring Harbor Press. (Expanded Editions, 1992, 2007).

Capra, F., & Luisi, P. L. 2014. *The Systems View of Life: A Unifying Vision.* Cambridge: Cambridge University Press.

Cardeña, E. 2015. "The Unbearable Fear of Psi: On Scientific Suppression in the 21st Century." *Journal of Scientific Exploration* 29: 601–20.

Cardeña, E., Palmer, J., & Marcusson-Clavertz, D. 2015. *Parapsychology: A Handbook for the 21st Century.* Jefferson, NC: McFarland.

Carey, B. 2011. "Journal's Paper on ESP Expected to Prompt Outrage." *New York Times* (January 6).

Chalmers, D. J. 1995. "Facing Up to the Problem of Consciousness." *Journal of Consciousness Studies* 2: 200–19.

Chalmers, D. J. 2002. *Philosophy of Mind: Classical and Contemporary Readings.* Oxford: Oxford University Press.

Chaoul, A. et al. 2018. "Randomized Trial of Tibetan Yoga in Patients with Breast Cancer Undergoing Chemotherapy." *Cancer* 124: 36–45.

Chawla, L. S., Akst, S., Junker, C., Jacobs, B., & Seneff, M. G. 2009. "Surges of Electroencephalogram Activity at the Time of Death: A Case Series." *Journal of Palliative Medicine* 12: 1095–1100.

Child, I. L. 1985. "Psychology and Anomalous Observations: The Question of ESP in Dreams." *American Psychologist* 40: 1219–30.

Cho, A. 2012. "The Discovery of the Higgs Boson." *Science* 338: 1524–25.

Cohen, L., Warneke, C., Fouladi, R. T., Rodriguez, M. A., & Chaoul-Reich, A. 2004. "Psychological Adjustment and Sleep Quality in a Randomized Trial of the Effects of a Tibetan Yoga Intervention in Patients with Lymphoma." *Cancer* 100: 2253–60.

Coleman, G., Jinpa, T., & Dorje, G. (Editors and Translator). 2006. *The Tibetan Book of the Dead: The Great Liberation by Hearing in the Intermediate States.* New York: Viking.

Cook, E. W., Greyson, B., & Stevenson, I. 1998. "Do Any Near-Death Experiences Provide Evidence for the Survival of Human Personality After Death? Relevant Features and Illustrative Case Reports." *Journal of Scientific Exploration* 12: 377–406.

Cuevas, B. J. 2003. *The Hidden History of the Tibetan Book of the Dead*. Oxford: Oxford University Press.

Dahaba, A. A. 2005. "Different Conditions That Could Result in the Bispectral Index Indicating an Incorrect Hypnotic State." *Anesthesia & Analgesia* 101: 765–73.

Dalai Lama, H. H. 1962. *My Land and My People: Autobiography of the Dalai Lama*. New York: McGraw-Hill.

——. 1998. *Freedom in Exile: The Autobiography of the Dalai Lama of Tibet* (Revised and Updated). London: Abacus. (Original, 1990).

——. 2005. *The Universe in a Single Atom: the Convergence of Science and Spirituality*. New York: Morgan Road.

Dalai Lama, H. H., & Ekman, P. 2008. *Emotional Awareness: Overcoming the Obstacles to Psychological Balance and Compassion*. New York: Henry Holt.

Darwin, C., & Wallace, A. 1858. "On the Tendency of Species to Form Varieties; and on the Perpetuation of Varieties and Species by Means of Selection." *Journal of Proceedings of the Linnean Society* 3: 45–62.

Das, N. N., & Gastaut, H. 1955. "Variations de l'activité électrique du cerveau, du coeur et des muscles squelletiques au cours de la meditation et de l'extase yogique." *Electroencephalography and Clinical Neurophysiology* Suppl. 6: 211–19.

Delbrück, M. 1986. *Mind from Matter? An Essay on Evolutionary Epistemology*. (Stent, G. S., Fischer, E. P., Golomb, S. W., Presti, D., & Seiler, H., Editors). Palo Alto, CA: Blackwell Scientific.

Devinsky, O., Feldmann, E., Burrowes, K., & Bromfield, E. 1989. "Autoscopic Phenomena with Seizures." *Archives of Neurology* 46: 1080–88.

du Prel, C. 1889. *The Philosophy of Mysticism* (C. C. Massey, Translator). London: George Redway. (Original in German, 1885).

Eccles, J. 1990. "A Unitary Hypothesis of Mind-Brain Interaction in the Cerebral Cortex." *Proceedings of the Royal Society of London, Series B, Biological Sciences* 240: 433–51.

Edelmann, J., & Bernet, W. 2007. "Setting Criteria for Ideal Reincarnation Research." *Journal of Consciousness Studies* 14: 92–101.

Ehrsson, H. H. 2007. "The Experimental Induction of Out-of-Body Experiences." *Science* 317: 1048.

Evans-Wentz, W. Y. 1960. *The Tibetan Book of the Dead; or The After-Death Experiences on the Bardo Plane, according to Lama Kazi Dawa-Samdup's English Rendering*. Oxford: Oxford University Press.

Fenwick, P., Lovelace, H., & Brayne, S. 2010. "Comfort for the Dying: Five Year Retrospective and One Year Prospective Studies of End of Life Experience." *Archives of Gerontology and Geriatrics* 51: 173–79.

Feyerabend, P. 2010. *Against Method.* (4th Edition). London: Verso. (1st Edition, 1975).

Feynman, R. P. 1965. *The Character of Physical Law.* Cambridge, MA: MIT Press.

Freeman, W. J. 2000. *How Brains Make Up Their Minds.* New York: Columbia University Press.

——. 2015. "Mechanism and Significance of Global Coherence in Scalp EEG." *Current Opinion in Neurobiology* 31: 199–205.

Fremantle, F., & Trungpa, C. (Editors and Translator). 1975. *The Tibetan Book of the Dead: The Great Liberation Through Hearing in the Bardo.* Boston: Shambhala.

Fuchs, C. A., Mermin, N. D., & Schack, R. 2014. "An Introduction to QBism with an Application to the Locality of Quantum Mechanics." *American Journal of Physics* 82: 749–54.

Fuchs, C. A., & Schack, R. 2013. "Quantum-Bayesian Coherence." *Reviews of Modern Physics* 85: 1693–715.

Gabbard, G.O., & Twemlow, S. W. 1984. *With the Eyes of the Mind: An Empirical Analysis of Out-of-Body States.* Westport, CT: Praeger.

Gardiner, B., Milner, R., & Morris, M. (Editors). 2008. "Survival of the Fittest: A Special Issue of *The Linnean* Celebrating the 150th Anniversary of the Darwin-Wallace Theory of Evolution." *The Linnean: Newsletter and Proceedings of The Linnean Society of London*, Special Issue 9.

Garfield, J. L. 1995. *The Fundamental Wisdom of the Middle Way: Nāgārjuna's Mūlamadhyamakakārikā.* Oxford: Oxford University Press.

——. 2012. "Ask Not What Buddhism Can Do for Cognitive Science; Ask What Cognitive Science Can Do for Buddhism." *Bulletin of Tibetology* 47: 15–30.

——. 2015. *Engaging Buddhism: Why It Matters to Philosophy.* Oxford: Oxford University Press.

Gauld, A. 1968. *Founders of Psychical Research.* New York: Schocken.

——. 1982. *Mediumship and Survival: A Century of Investigations.* London: Heinemann.

Gefter, A., & Hoffman, D. D. 2016. "The Evolutionary Argument Against Reality." *Quanta Magazine* (April 21).

Goldberg, P. 2010. *American Veda: From Emerson and the Beatles to Yoga and Meditation—How Indian Spirituality Changed the West.* New York: Harmony.

Goleman, D. J., & Schwartz G. E. 1976. "Meditation as an Intervention in Stress Reactivity." *Journal of Consulting and Clinical Psychology* 44: 456–66.

Goncharova, I. I., McFarland, D. J., Vaughan, T. M., & Wolpaw, J. R. 2003. "EMG Contamination of EEG: Spectral and Topographical Characteristics." *Clinical Neurophysiology* 114: 1580–93.

Gould, S. J. 1997. "Nonoverlapping Magisteria." *Natural History* 106(March): 16–22.

Gould, S. J. 1999. *Rock of Ages: Science and Religion in the Fullness of Life.* New York: Ballantine.

Gray, D. B. 2007. *The Chakrasamvara Tantra (The Discourse of Sri Heruka).* New York: America Institute of Buddhist Studies.

Graziano, M. S. A. 2013. *Consciousness and the Social Brain.* Oxford: Oxford University Press.

Greene, B. 1999. *The Elegant Universe: Superstrings, Hidden Dimensions, and the Quest for the Ultimate Theory.* New York: Norton.

Greyson, B. 1981. "Near Death Experiences and Attempted Suicide." *Suicide & Life Threatening Behavior* 11: 10–16.

——. 1992. "Near-Death Experiences and Anti-Suicidal Attitudes." *Omega* 26: 81–9.

——. 1998. "The Incidence of Near-Death Experiences." *Medicine & Psychiatry* 1: 92–9.

——. 2003. "Incidence and Correlates of Near-Death Experiences in a Cardiac Care Unit." *General Hospital Psychiatry* 25: 269–76.

——. 2007. "Comment on 'Does Paranormal Perception Occur in Near-Death Experiences?'" *Journal of Near-Death Studies* 25: 237–44.

——. 2010. "Seeing Dead People Not Known to Have Died: 'Peak in Darien' Experiences." *Anthropology & Humanism* 35: 159–71.

Greyson, B., Fountain, N. B., Derr. L. L., & Broshek, D. K. 2014. "Out-Of-Body Experiences Associated With Seizures." *Frontiers in Human Neuroscience* 8: 65. https://doi.org/10.3389/fnhum.2014.00065

Greyson, B., Kelly, E.F., & Dunseath, W. J. R. 2013. "Surge of Neurophysiological Activity in the Dying Brain." *Proceedings of the National Academy of Sciences USA* 110: E4405.

Griffiths, R. R., Richards, W. A., McCann, U., & Jesse, R. 2006. "Psilocybin Can Occasion Mystical-Type Experiences Having Substantial and Sustained Personal Meaning and Spiritual Significance." *Psychopharmacology* 187: 268–83.

Grosso, M. 2016. *The Man Who Could Fly: St. Joseph of Copertino and the Mystery of Levitation.* Lanham, MD: Rowman & Littlefield.

———. 2017. *Wings of Ecstasy: Domenico Bernini's Vita of St. Joseph of Copertino (1722).* (Translated by C. Clough; abridged with commentary by M. Grosso). CreateSpace Independent Publishing Platform.

Gurney, E., Myers, F. W. H., & Podmore, F. 1886. *Phantasms of the Living* (2 vols.). London: Trübner.

Haig, S. 2007. "The Brain: The Power of Hope." *Time Magazine* 169(5): 118–19 (January 29).

Hansen, G. P. 2001. *The Trickster and the Paranormal.* Bloomington, IN: Xlibris.

Haraldsson, E. 1995. "Personality and Abilities of Children Claiming Previous-Life Memories." *Journal of Nervous and Mental Disease* 183: 445–51.

———. 2003. "Children who Speak of Past-Life Experiences: Is There a Psychological Explanation?" *Psychology and Psychotherapy: Theory, Research and Practice* 76: 55–67.

———. 2008. "Persistence of Past-Life Memories. A Study of Persons Who Claimed in Their Childhood to Remember a Past Life." *Journal of Scientific Exploration* 22: 385–93.

Haraldsson, E., Fowler, P. C., & Periyannanpillai, V. 2000. "Psychological Characteristics of Children Who Speak of a Previous Life: A Further Field Study in Sri Lanka." *Transcultural Psychiatry* 37: 525–44.

Harner, M. J. (Editor). 1973. *Hallucinogens and Shamanism.* Oxford: Oxford University Press.

Harner, M. 2013. *Cave and Cosmos: Shamanic Encounters with Another Reality.* Berkeley, CA: North Atlantic.

Harrington, A., & Zajonc, A. (Editors). 2006. *The Dalai Lama at MIT.* Cambridge, MA: Harvard University Press.

Harrison, P. 2015. *The Territories of Science and Religion.* Chicago: University of Chicago Press.

Hart, H., & Hart, E. B. 1933. "Visions and Apparitions Collectively and Reciprocally Perceived." *Proceedings of the Society for Psychical Research* 41: 205–49.

Hasenkamp, W., & White, J. R. (Editors). 2017. *The Monastery and the Microscope: Conversations with the Dalai Lama on Mind, Mindfulness, and the Nature of Reality.* New Haven, CT: Yale University Press.

Hawking, S., & Mlodinow, L. 2010. *The Grand Design.* New York: Bantam.

Hayward, J. W., & Varela, F. J. (Editors). 1992. *Gentle Bridges: Conversations with the Dalai Lama on the Sciences of Mind.* Boston: Shambhala.

Heim, A. 1892. "Notizen über den Tod Absturz [Remarks on Fatal Falls]." *Jahrbuch der Schweitzerischen Alpclub [Yearbook of the Swiss Alpine Club]* 27: 327–37.

Holck, F. H. 1978. "Life Revisited (Parallels in Death Experiences)." *Omega* 9: 1–11.

Holden, J. M., Greyson, B., & James, D. (Editors). 2009. *The Handbook of Near-Death Experiences: Thirty Years of Investigation.* Westport, CT: Praeger.

Holden, J. M., Long, J., & MacLurg, J. 2006. "Out-Of-Body Experiences: All in the Brain?" *Journal of Near-Death Studies* 25: 99–107.

Hölzel, B. K., Lazar, S. W., Gard, T., Schuman-Olivier, Z., Vago, D. R., & Ott, U. 2011. "How Does Mindfulness Meditation Work? Proposing Mechanisms of Action From a Conceptual and Neural Perspective." *Perspectives on Psychological Science* 6: 537–59.

Hook, E. B. (Editor). 2002. *Prematurity in Scientific Discovery.* Berkeley: University of California Press.

Houshmand, Z., Livingston, R. B., & Wallace, B. A. (Editors). 1999. *Consciousness at the Crossroads: Conversations with the Dalai Lama on Brain Science and Buddhism.* Ithaca, NY: Snow Lion.

Huxley, A. 1954. *The Doors of Perception.* New York: Harper & Row.

Jacobs, T. L., Epel, E. S., Lin, J., Blackburn, E. H., Wolkowitz, O. M., Bridwell, D. A., Zanesco, A. P., Aichele, A. R., Sahdra, B. K., MacLean, K. A., King, B. G., Shaver, P. R., Rodenberg, E. L., Ferrer, E., Wallace, B. A., & Saron, C. D. 2011. "Intensive Meditation Training, Immune Cell Telomerase Activity, and Psychological Mediators." *Psychoneuroendocrinology* 36: 664–81.

James, W. 1890. *The Principles of Psychology.* New York: Henry Holt.

———. 1892. *Psychology: The Briefer Course.* New York: Henry Holt.

———. 1900. *Human Immortality: Two Supposed Objections to the Doctrine* (2nd Edition). Boston: Houghton Mifflin. (1st Edition, 1898).

——. 1902. *The Varieties of Religious Experience: A Study in Human Nature.* London: Longmans, Green.

——. 1909. *A Pluralistic Universe: Hibbert Lectures to Manchester College on the Present Situation in Philosophy.* New York: Longmans.

——. 1986. *The Works of William James: Essays in Psychical Research.* Cambridge, MA: Harvard University Press.

Jinpa, T. 2010. "Buddhism and Science: How Far Can the Dialogue Proceed?" *Zygon* 45: 871–82.

Judson, H. F. 1979. *The Eighth Day of Creation: Makers of the Revolution in Biology.* New York: Simon and Schuster.

Kabat-Zinn, J. 1982. "An Outpatient Program in Behavioral Medicine for Chronic Pain Patients Based on the Practice of Mindfulness Meditation: Theoretical Considerations and Preliminary Results." *General Hospital Psychiatry* 4: 33–47.

Kakar, S., & Kripal, J. J. (Editors). 2012. *Seriously Strange: Thinking Anew about Psychical Experiences.* New Delhi, India: Penguin Viking.

Kaliman, P., Álvarez-López, M. J., Cosín-Tomás, M., Rosenkranz, M. A., Lutz, A., & Davidson, R. J. 2014. "Rapid Changes in Histone Deacetylases and Inflammatory Gene Expression in Expert Meditators." *Psychoneuroendocrinology* 40: 96–107.

Kanthamani, H., & Kelly, E. F. 1975. "Card Experiments with a Special Subject. II. The Shuffle Method." *Journal of Parapsychology* 39: 206–21.

Kelly, E. F. 1982. "On Grouping of Hits in Some Exceptional Psi Performers." *Journal of the American Society for Psychical Research* 76: 101–42.

Kelly, E. F., Crabtree, A., & Marshall, P. (Editors). 2015. *Beyond Physicalism: Toward Reconciliation of Science and Spirituality.* Lanham, MD: Rowman & Littlefield.

Kelly, E. F., & Kanthamani, B. K. (H). 1972. "A Subject's Efforts Toward Voluntary Control." *Journal of Parapsychology* 36: 185–97.

Kelly, E. F., Kanthamani, H., Child, I. L., & Young, F. W. 1975. "On the Relation Between Visual and ESP Confusion Structures in an Exceptional ESP Subject." *Journal of the American Society for Psychical Research* 69: 1–31.

Kelly, E. F., Kelly, E. W., Crabtree, A., Gauld, A., Grosso, M., & Greyson, B. 2007. *Irreducible Mind: Toward A Psychology for the 21st Century.* Lanham, MD: Rowman & Littlefield.

Kelly, E. F., & Locke, R. L. 2009. *Altered States of Consciousness and Psi: An Historical Survey and Research Prospectus* (Parapsychological Monographs No. 18). New York: Parapsychology Foundation. (Original, 1981).

Kelly, E. W. 2010a. "Near-Death Experiences with Reports of Meeting Deceased People." *Death Studies* 25: 229–49.

——. 2010b. "Some Directions for Research on Mediumship." *Journal of Scientific Exploration* 24: 247–82.

Kelly, E. W. (Editor). 2013. *Science, the Self, and Survival after Death: Selected Writings of Ian Stevenson, M.D.* Lanham, MD: Rowman & Littlefield.

Kelly, E. W., & Alvarado, C. S. 2005. "Images in Psychiatry: Frederic William Henry Myers." *American Journal of Psychiatry* 162: 34.

Kelly, E. W., & Arcangel, D. 2011. "An Investigation of Mediums Who Claim to Give Information About Deceased Persons." *Journal of Nervous and Mental Disease* 199: 11–7.

Kennedy, D. et al. 2005. "What Don't We Know?" *Science* 309: 75–102.

King, L. S. 1975. "Reincarnation." *Journal of the American Medical Association* 234: 978.

Knapp, A. 2012. "How Much Does it Cost to Find a Higgs Boson?" *Forbes* (July 5).

Koch, C. 2004. *The Quest for Consciousness: A Neurobiological Approach.* Englewood, CO: Roberts and Company.

——. 2012. *Consciousness: Confessions of a Romantic Reductionist.* Cambridge, MA: MIT Press.

Koshland, D.E., Jr. 2002. "The Seven Pillars of Life." *Science*, 295: 2215–16.

Kripal, J. J. 2007. *Esalen: America and the Religion of No Religion.* Chicago: University of Chicago Press.

——. 2010. *Authors of the Impossible: The Paranormal and the Sacred.* Chicago: University of Chicago Press.

——. 2011. *Mutants and Mystics: Science Fiction, Superhero Comics, and the Paranormal.* Chicago, IL: University of Chicago Press.

——. 2014. *Comparing Religions: Coming to Terms.* Chichester, UK: Wiley Blackwell.

——. 2016. "Introduction: Reimagining the Super in the Study of Religion." *Religion: Super Religion.* (Macmillan Interdisciplinary Handbook), xv–xlviii. Boston: Cengage.

Kübler-Ross, E. 1983. *On Children and Death*. New York: Macmillan.

Kuhn, T. S. 1957. *The Copernican Revolution: Planetary Astronomy in the Development of Western Thought*. Cambridge, MA: Harvard University Press.

——. 1996. *The Structure of Scientific Revolutions* (3rd Edition). Chicago: University of Chicago Press. (1st Edition, 1962).

Kumar, M. 2008. *Quantum: Einstein, Bohr and the Great Debate about the Nature of Reality*. New York: Norton.

Leal, I., Engebretson, J., Cohen, L., Fernandez-Esquer, M. E., Lopez, G., Wangyal, T., & Chaoul, A. 2018. "An Exploration of the Effects of Tibetan Yoga on Patients' Psychological Well-Being and Experience of Lymphoma: An Experimental Embedded Mixed Methods Study." *Journal of Mixed Methods Research* 12: 31–54.

Leal, I., Engebretson, J., Cohen, L., Rodriguez, A., Wangyal, T., Lopez, G., & Chaoul, A. 2015. "Experiences of Paradox: A Qualitative Analysis of Living with Cancer Using a Framework Approach." *Psycho-Oncology* 24: 138–46.

Leary, M. 2011. "Why Are (Some) Scientists So Opposed to Parapsychology?" *Explore* 7: 275–77.

Leininger, B., Leininger, A., & Gross, K. 2009. *Soul Survivor: The Reincarnation of a World War II Fighter Pilot*. New York: Grand Central.

Lenggenhager, B., Tadi, T., Metzinger, T., & Blanke, O. 2007. "Video ergo Sum: Manipulating Bodily Self-Consciousness." *Science* 317: 1096–99.

Lerner, E. 2004. "Bucking the Big Bang." *New Scientist* 182(2448), 20 (May 22).

Levin, J. 1983. "Materialism and Qualia: The Explanatory Gap." *Pacific Philosophical Quarterly* 64: 354–61.

Lopez, D. S., Jr. 2008. *Buddhism and Science: A Guide for the Perplexed*. Chicago: University of Chicago Press.

——. 2010. "The Future of the Buddhist Past: A Response to the Readers." *Zygon* 45: 883–96.

Lopez, D. S., Jr. 2011. *The Tibetan Book of the Dead: A Biography*. Princeton, NJ: Princeton University Press.

Ludwig, D. S., & Kabat-Zinn, J. 2008. "Mindfulness in Medicine." *Journal of the American Medical Association* 300: 1350–52.

Luisi, P. L., & Houshmand, Z. 2009. *Mind and Life: Discussions with the Dalai Lama on the Nature of Reality*. New York: Columbia University Press.

Luke, D. 2017. *Otherworlds: Psychedelics and Exceptional Human Experience.* London: Muswell Hill.

Lutz, A., Dunne, J. D., & Davidson, R. J. 2007. "Meditation and the Neuroscience of Consciousness: An Introduction." *The Cambridge Handbook of Consciousness* (Zelazo, P. D., Moscovitch, M., & Thompson, E., Editors), 499–554. Cambridge: Cambridge University Press.

Lutz, A., Greischar, L. L., Rawlings, N. B., Ricard, M., & Davidson, R. J. 2004. "Long-Term Meditators Self-Induce High-Amplitude Gamma Synchrony During Mental Practice." *Proceedings of the National Academy of Sciences USA* 101: 16369–73.

Lutz, A., Jha, A. P., Dunne, J. D., & Saron, C. D. 2015. "Investigating the Phenomenological Matrix of Mindfulness-Related Practices From a Neurocognitive Perspective." *American Psychologist* 70: 632–58.

MacLean, K. A., Ferrer, E., Aichele, S. R., Bridwell, D. A., Zanesco, A. P., Jacobs, T. L., King, B. G., Rodenberg, E. L., Sahdra, B. K., Shaver, P. R., Wallace, B. A., Mangun, G. R., & Saron, C. D. 2010. "Intensive Meditation Training Improves Perceptual Discrimination and Sustained Attention." *Psychological Science* 21: 829–39.

Mark, J. T., Marion, B. B., & Hoffman, D. D. 2010. "Natural Selection and Veridical Perceptions." *Journal of Theoretical Biology* 266: 504–15.

Marshall, P. 2005. *Mystical Encounters with the Natural World: Experiences and Explanations.* Oxford: Oxford University Press.

Mathison, M. 2011. "The Time Has Come for Me to Retire [an interview with the Dalai Lama]." *Rolling Stone* 1136: 54–59, 82 (4 August).

Maturana, H. R., & Varela, F. J. 1980. *Autopoiesis and Cognition: The Realization of the Living.* Dordrecht, NL: D. Reidel.

Mauskopf, S. H., & McVaugh, M. R. 1980. *The Elusive Science: Origins of Experimental Psychical Research.* Baltimore: The Johns Hopkins University Press.

Mayer, E. L. 2007. *Extraordinary Knowing: Science, Skepticism, and the Inexplicable Powers of the Human Mind.* New York: Bantam.

Mermin, N. D. 2014. "QBism Puts the Scientist Back Into Science." *Nature* 507: 421–23.

Meyers, W., Thurman, R., & Burbank, M. G. 2016. *Man of Peace: The Illustrated Life Story of the Dalai Lama of Tibet.* New York: Tibet House US.

Milbury, K., Chaoul, A., Biegler, K., Wangyal, T., Spelman, A., Meyers, C. A., Arun, B., Palmer, J. L., Taylor, J., & Cohen, L. 2013. "Tibetan Sound

Meditation for Cognitive Dysfunction: Results of a Randomized Controlled Pilot Trial." *Psycho-Oncology* 22: 2354–63.

Milbury, K., Chaoul, A., Engle, R., Liao, Z., Yang, C., Carmack, C., Shannon, V., Spelman, A., Wangyal, T., & Cohen, L. 2015. "Couple-Based Tibetan Yoga Program for Cancer Patients and Their Caregivers." *Psycho-Oncology* 24: 117–20.

Miller, G. 2005. "What Is the Biological Basis of Consciousness?" *Science* 309: 79.

Mills, A., Haraldsson, E., & Keil, H. H. J. 1994. "Replication Studies of Cases Suggestive of Reincarnation by Three Independent Investigators." *Journal of the American Society for Psychical Research* 88: 207–19.

Moody, R. A. 2001. *Life After Life* (25th Anniversary Edition). New York: Bantam Harper. (Original, 1975).

Murphy, G., & Ballou, R. O. (Editors). 1960. *William James on Psychical Research*. New York: Viking.

Myers, F. W. H. 1903. *Human Personality and Its Survival of Bodily Death* (two volumes). London: Longmans, Green.

Myles, P. S., & Cairo, S. 2004. "Artifact in the Bispectral Index in a Patient with Severe Ischemic Brain Injury." *Anesthesia & Analgesia* 98: 706–07.

Nagel, T. 1974. "What Is Like To Be a Bat?" *Philosophical Review* 83: 435–50.

——. 1986. *The View From Nowhere*. Oxford: Oxford University Press.

——. 2012. *Mind & Cosmos: Why the Materialist Neo-Darwinian Conception of Nature is Almost Certainly False*. Oxford: Oxford University Press.

——. 2017. "Is Consciousness an Illusion?" *The New York Review of Books*, 64(4) (March 9).

Nahm, M. 2011. "The *Tibetan Book of the Dead*: Its History and Controversial Aspects of its Contents." *Journal of Near-Death Studies* 29: 373–98.

Nahm, M., Greyson, B., Kelly, E. W., & Haraldsson, E. 2012. "Terminal Lucidity: A Review and a Case Collection." *Archives of Gerontology and Geriatrics* 55: 138–42.

Noyes, R., & Kletti, R. 1972. "The Experience of Dying From Falls." *Omega* 3: 45–52.

Open Science Collaboration. 2015. "Estimating the Reproducibility of Psychological Science." *Science* 349: 943.

Osis, K., & Haraldsson, E. 1997. *At the Hour of Death* (3rd Edition). Norwalk, CT: Hastings. (1st Edition, 1977).

Osto, D. 2016. *Altered States: Buddhism and Psychedelic Spirituality in America.* New York: Columbia University Press.

Owens, J. E., Cook, E. W., & Stevenson, I. 1990. "Features of 'Near-Death Experience' in Relation to Whether or Not Patients Were Near Death." *Lancet* 336: 1175–77.

Parnia, S. 2013. *Erasing Death: The Science That Is Rewriting the Boundaries Between Life and Death.* New York: Harper.

Parnia, S., & Fenwick, P. 2002. "Near Death Experiences in Cardiac Arrest: Visions of a Dying Brain or Visions of a New Science of Consciousness." *Resuscitation* 52: 5–11.

Parnia, S. et al. 2014. "AWARE—AWAreness During REsuscitation—A Prospective Study." *Resuscitation* 85: 1799–1805.

Parnia, S., Waller, D. G., Yeates, Y., & Fenwick, P. 2001. "A qualitative and Quantitative Study of the Incidence, Features and Aetiology of Near Death Experiences in Cardiac Arrest Survivors." *Resuscitation* 48: 149–56.

Pearson, H. 2008. "'Virophage' Suggests Viruses Are Alive." *Nature* 454: 677.

Penrose, R. 1989. *The Emperor's New Mind: Concerning Computers, Minds, and The Laws of Physics.* Oxford: Oxford University Press.

Penrose, R. 1997. *The Large, the Small and the Human Mind.* Cambridge, UK: Cambridge University Press.

Penrose, R. 2004. *The Road to Reality: A Complete Guide to the Laws of the Universe.* New York: Alfred Knopf.

Perutz, M. F. 1987. "Physics and the Riddle of Life." *Nature* 326: 555–58.

Planck, M. 1933. *Where is Science Going?* (Translated and edited by J. Murphy). London: George Allen & Unwin Ltd.

Pliny the Elder. 1942. *Natural History, Vol. 2, Book 7* (H. Rackham, Translator). Cambridge, MA: Harvard University Press.

Pollan, M. 2018. *How to Change Your Mind: What the New Science of Psychedelics Teaches Us About Consciousness, Dying, Addiction, Depression, and Transcendence.* London: Penguin.

Ponlop, D. 2006. *Mind Beyond Death.* Ithaca, NY: Snow Lion.

Pratt, J. G., Rhine, J. B., Smith, B. M., Stuart, C. E., & Greenwood, J. A. 1940. *Extra-Sensory Perception After Sixty Years: A Critical Appraisal of the Research in Extra-Sensory Perception.* New York: Henry Holt.

Presti, D. E. 2016. *Foundational Concepts in Neuroscience: A Brain-Mind Odyssey.* New York: Norton.

Radin, D. 1997. *The Conscious Universe: The Scientific Truth of Psychic Phenomena*. New York: Harper.

———. 2006. *Entangled Minds: Extrasensory Experiences in a Quantum Reality*. New York: Simon & Schuster.

———. 2013. *Supernormal: Science, Yoga, and the Evidence for Extraordinary Psychic Abilities*. New York: Random House.

Randall, L. 2005. *Warped Passages: Unraveling the Mysteries of the Universe's Hidden Dimensions*. New York: HarperCollins.

Rao, K. R. (Editor). 2010. *Yoga and Parapsychology: Empirical Research and Theoretical Essays*. Delhi, India: Motilal Banarsidass Publishers.

Rao, K. R., Paranjape, A. C., & Dalai, A. K. (Editors). 2008. *Handbook of Indian Psychology*. New Delhi, India: Cambridge University Press India.

Ricard, M., & Singer, W. 2017. *Beyond the Self: Conversations Between Buddhism and Neuroscience*. Cambridge, MA: MIT Press.

Ricard, M., & Thuan, T. X. 2001. *The Quantum and the Lotus: A Journey to the Frontiers Where Science and Buddhism Meet*. New York: Three Rivers Press. (Original in French as *L'Infini dans la Paume de la Main*, 2000).

Richards, W. 2015. *Sacred Knowledge: Psychedelics and Religious Experiences*. New York: Columbia University Press.

Ring, K., & Cooper, S. 1999. *Mindsight: Near-Death and Out-of-Body Experiences in the Blind*. Palo Alto, CA: Institute of Transpersonal Psychology.

Roberts, G., & Owen, J. 1988. "The Near-Death Experience." *British Journal of Psychiatry* 153: 607–17.

Rosch, E. 1999. "Is Wisdom In the Brain?" *Psychological Science* 10: 222–24.

Rosenblum, B., & Kuttner, F. 2011. *Quantum Enigma: Physics Encounters Consciousness* (2nd Edition). Oxford: Oxford University Press. (1st Edition, 2006).

Ruse, M. (Editor). 2009. *Philosophy After Darwin: Classic and Contemporary Readings*. Princeton, NJ: Princeton University Press.

Rush, B. 1812. *Mental Inquiries and Observations upon Diseases of the Mind*. Philadelphia: Kimber & Richardson.

Sabom, M. B. 1982. *Recollections of Death: A Medical Investigation*. New York: Harper.

———. 1998. *Light And Death: One Doctor's Fascinating Account Of Near-Death Experiences*. Grand Rapids, MI: Zondervan.

Sagan, C. 1996. *The Demon-Haunted World: Science as a Candle in the Dark.* New York: Random House.

Sager, B. 2012. *Beyond the Robe: Science for Monks and All It Reveals about Tibetan Monks and Nuns.* New York: powerHouse.

Sartori, P. 2008. *The Near-Death Experiences of Hospitalized Intensive Care Patients: A Five Year Clinical Study.* Lewiston, NY: Edwin Mellen.

Sawyer, D. 2014. *Huston Smith: Wisdomkeeper—Living the World's Religions, The Authorized Biography of a 21st Century Spiritual Giant.* Louisville, KY: Fons Vitae.

Schneider, S., & Velmans, M. (Editors). 2017. *Blackwell Companion to Consciousness* (2nd Edition). Chichester, UK: Wiley.

Schooler, J. W. 2014. "Turning the Lens of Science on Itself: Verbal Overshadowing, Replication, and Metascience." *Perspectives on Psychological Science* 9: 579–84.

Schouten, S. A., & Stevenson, I. 1998. "Does the Socio-Psychological Hypothesis Explain Cases of the Reincarnation Type?" *Journal of Nervous and Mental Disease* 186: 504–6.

Schrödinger, E. 1944. *What Is Life?* Cambridge: Cambridge University Press.

Seife, C. 2005. "What Is the Universe Made Of?" *Science* 309: 78–9.

Shanon, B. 2002. *The Antipodes of the Mind: Charting the Phenomenology of the Ayahuasca Experience.* Oxford: Oxford University Press.

Sharf, R. H. 2018. "Knowing Blue: Early Buddhist Accounts of Non-Conceptual Sense Perception." *Philosophy East and West.* (In press). doi: 10.1353/pew.0.0126

Sharma, P., & Tucker, J. B. 2004. "Cases of the Reincarnation Type With Memories From the Intermission Between Lives." *Journal of Near-Death Studies* 23: 101–18.

Sharp, K. C. 1995. *After the Light: What I discovered on the Other Side of Life that can Change your World.* New York: William Morrow.

Shroder, T. 1999. *Old Souls: The Scientific Evidence for Past Lives.* New York: Simon & Schuster.

Shushan, G. 2009. *Conceptions of the Afterlife in Early Civilizations: Universalism, Constructivism, and Near-Death Experience.* London: Continuum.

Smythies, J. R. (Editor). 1967. *Science and ESP.* London: Routledge & Kegan Paul.

Smythies, J. 2012. "Consciousness and Higher Dimensions of Space." *Journal of Consciousness Studies* 19: 224–32.

Sogyal Rinpoche. 1992. *The Tibetan Book of Living and Dying*. New York: HarperCollins.

Sonam, N., Gyaltsen, P. L., & Gyatso, K. (Editors). 1974. *The Vast Treasury of Profound Space*, 321–46. Thobgyal Sonam, India: Tibetan Bonpo Monastic Center.

Stapp, H. P. 1996. "The Hard Problem: A Quantum Approach." *Journal of Consciousness Studies* 3: 194–210.

Stapp, H. P. 2009. *Mind, Matter and Quantum Mechanics* (3rd Edition). Berlin: Springer-Verlag.

Stapp, H. P. 2011. *Mindful Universe: Quantum Mechanics and the Participating Observer* (2nd Edition). Berlin: Springer-Verlag.

Steiger, B., & Steiger, S. H. 1995. *Children of the Light: The Startling And Inspiring Truth About Children's Near-Death Experiences And How They Illumine The Beyond*. New York: Signet Penguin.

Stent, G. S. 1968. "That Was the Molecular Biology That Was." *Science* 160: 390–5.

Stent, G. S. 1972. "Prematurity and Uniqueness in Scientific Discovery." *Scientific American* 227(6): 84–93.

Stevenson, I. 1960. "The Evidence for Survival From Claimed Memories of Former Incarnations." *Journal of the American Society for Psychical Research* 54: 51–71, 95–117.

———. 1974. *Twenty Cases Suggestive of Reincarnation* (Revised Edition; Original, 1966). Charlottesville: University of Virginia Press.

———. 1975. *Cases of the Reincarnation Type, Volume I: Ten Cases in India*. Charlottesville: University of Virginia Press.

———. 1977. *Cases of the Reincarnation Type, Volume II: Ten Cases in Sri Lanka*. Charlottesville: University of Virginia Press.

———. 1980. *Cases of the Reincarnation Type, Volume III: Twelve Cases in Lebanon and Turkey*. Charlottesville: University of Virginia Press.

———. 1983. *Cases of the Reincarnation Type, Volume IV: Twelve Cases in Thailand and Burma*. Charlottesville: University of Virginia Press.

———. 1986. "Characteristics of Cases of the Reincarnation Type Among the Igbo of Nigeria." *Journal of Asian and African Studies* 21: 204–16.

———. 1990. "Phobias in Children Who Claim to Remember Previous Lives." *Journal of Scientific Exploration* 4: 243–54.

———. 1995. "Six Modern Apparitional Experiences." *Journal of Scientific Exploration* 9: 351–66.

——. 1997a. *Reincarnation and Biology: A Contribution to the Etiology of Birthmarks and Birth Defects.* Westport, CT: Praeger.

——. 1997b. *Where Reincarnation and Biology Intersect.* Westport, CT: Praeger.

——. 2000. "Unusual Play in Young Children Who Claim to Remember Previous Lives." *Journal of Scientific Exploration* 14: 557–70.

——. 2001. *Children Who Remember Previous Lives: A Question of Reincarnation* (Revised Edition). Jefferson, NC: McFarland (Original, 1987).

Stevenson, I., & Cook, E. W. 1995. "Involuntary Memories During Severe Physical Illness or Injury. *Journal of Nervous & Mental Disease* 183: 452–58.

Stevenson, I. & Keil, J. 2000. "The Stability of Assessments of Paranormal Connections in Reincarnation-Type Cases." *Journal of Scientific Exploration* 14: 365–82.

Stevenson, I., & Keil, J. 2005. "Children of Myanmar Who Behave Like Japanese Soldiers: A Possible Third Element in Personality." *Journal of Scientific Exploration* 19: 171–83.

Strieber, W., & Kripal, J. J. 2016. *The Super Natural: A New Vision of the Unexplained.* New York: Jeremy Tarcher/Penguin.

Sudduth, M. 2016. *A Philosophical Critique of Empirical Arguments for Postmortem Survival.* London: Palgrave Macmillan.

Sutherland, C. 1995. *Children of the Light: Near-Death Experiences of Children.* New York: Bantam.

Symonds, N. 1987. Schrödinger and *What is Life? Nature* 327: 663–4.

Taimni, I. K. 1961. *The Science of Yoga: The Yoga Sutras of Patanjali.* Chennai, India: Theosophical Publishing House.

Tang, Y. Y., Lu, Q., Fan, M., Yang, Y., & Posner, M. I. 2012. "Mechanisms of White Matter Changes Induced by Meditation." *Proceedings of the National Academy of Sciences USA* 109: 10570–74.

Tang, Y. Y., Ma, Y., Wang, J., Fan, Y., Feng, S., Lu, Q., Yu, Q., Sui, D., Rothbart, M. K., Fan, M., & Posner, M. I. 2007. "Short-Term Meditation Training Improves Attention and Self-Regulation." *Proceedings of the National Academy of Sciences USA* 104: 17152–56.

Tart, C. T. 2009. *The End of Materialism: How Evidence of the Paranormal is Bringing Science and Spirit Together.* Oakland, CA: New Harbinger.

Thompson, E. 2007. *Mind in Life: Biology, Phenomenology, and the Sciences of Mind.* Cambridge, MA: Harvard University Press.

——. 2015. *Waking, Dreaming, Being: Self and Consciousness in Neuroscience, Meditation, and Philosophy.* New York: Columbia University Press.

Thurman, R. A. F. 1998. *The Tibetan Book of the Dead.* New York: Bantam.

Tononi, G. (2008). "Consciousness As Integrated Information: A Provisional Manifesto." *Biological Bulletin* 215: 216–42.

Tononi, G., Boly, M., Massimini, M., & Koch, C. 2016. "Integrated Information Theory: From Consciousness to Its Physical Substrate." *Nature Reviews Neuroscience* 17: 450–61.

Tucker, J. B. 2000. "A Scale to Measure the Strength of Children's Claims of Previous Lives: Methodology and Initial Findings." *Journal of Scientific Exploration* 14: 571–81.

——. 2005. *Life Before Life: A Scientific Investigation of Children's Memories of Previous Lives.* New York: St. Martin's.

——. 2007. "Children Who Claim to Remember Previous Lives: Past, Present, and Future Research." *Journal of Scientific Exploration* 21: 543–52.

——. 2013. *Return to Life: Extraordinary Cases of Children Who Remember Past Lives.* New York: St. Martin's.

——. 2016. "The Case of James Leininger: An American Case of the Reincarnation Type." *EXPLORE* 12: 200–07.

Tucker, J. B., & Keil, H. H. J. 2001. "Can Cultural Beliefs Cause a Gender Identity Disorder?" *Journal of Psychology & Human Sexuality* 13: 21–30.

Tucker, J. B., & Keil, H. H. J. 2013. "Experimental Birthmarks: New Cases of an Asian Practice." *Journal of Scientific Exploration* 27: 269–82.

Tucker, J. B., & Nidiffer, F. D. 2014. "Psychological Evaluation of American Children Who Report Memories of Previous Lives." *Journal of Scientific Exploration* 23: 583–94.

Tyrrell, G. N. M. 1953. *Apparitions.* London: Duckworth. (Original, 1943).

Ullman, M., Krippner, S., & Vaughan, A. 1973. *Dream Telepathy: An Experimental Odyssey.* New York: Macmillan.

——. 2003. *Dream Telepathy: Experiments in Nocturnal Extrasensory Perception.* Newburyport, MA: Hampton Roads.

van Dongen, H., Gerding, H., & Sneller, R. 2014. *Wild Beasts of the Philosophical Desert: Philosophers on Telepathy and Other Exceptional Experiences.* Newcastle upon Tyne, UK: Cambridge Scholars Publishing.

van Lommel, P. 2004. "About the Continuity of Our Consciousness." *Advances in Experimental Medicine and Biology* 550: 115–32.

———. 2010. *Consciousness Beyond Life: The Science of the Near-Death Experience.* New York: Harper.

———. 2011. "Near-Death Experiences: The Experience of the Self as Real and Not as an Illusion." *Annals of the New York Academy of Sciences* 1234: 19–28.

van Lommel, P., Wees, R., Meyers, V., & Elfferich, I. 2001. "Near-Death Experience in Survivors of Cardiac Arrest: A Prospective Study in the Netherlands. *Lancet* 358: 2039–45.

Varela, F. J. (Editor). 1997. *Sleeping, Dreaming, and Dying: An Exploration of Consciousness with the Dalai Lama.* Boston: Wisdom Publications.

Varela, F. J., Thompson, E., & Rosch, E. 2016. *The Embodied Mind: Cognitive Science and Human Experience.* Cambridge, MA: MIT Press. (Original, 1991; Revised edition has same text as original, with new forward and introductions).

Velmans, M. 2009. *Understanding Consciousness* (2nd Edition). London: Routledge.

Voss, U., Holzmann, R., Hobson, A., Paulus, W., Koppehele-Grosssel, J., Klimke, A., & Nitsche, M. A. 2014. "Induction of Self Awareness in Dreams Through Frontal Low Current Stimulation of Gamma Activity." *Nature Neuroscience* 17: 810–12.

Wackermann, J., Pütz, P., & Allefeld, C. 2008. "Ganzfeld-Induced Hallucinatory Experience, Its Phenomenology and Cerebral Electrophysiology." *Cortex* 44: 1364–78.

Wachelder, E. M., Moulaert, V. R., van Heugten, C., Verbunt, J. A., Bekkers, S. C., & Wade, D.T. 2009. "Life After Survival: Long-Term Daily Functioning and Quality of Life After an Out-Of-Hospital Cardiac Arrest." *Resuscitation* 80: 517–22.

Wallace, B. A. 1996. *Choosing Reality: A Buddhist View of Physics and the Mind.* Ithaca, NY: Snow Lion.

———. 1999. "The Buddhist Tradition of Samatha: Methods for Refining and Examining Consciousness." *Journal of Consciousness Studies* 6: 175–87.

———. 2000. *The Taboo of Subjectivity: Toward a New Science of Consciousness.* Oxford: Oxford University Press.

Wallace, B. A. (Editor). 2003. *Buddhism and Science: Breaking New Ground.* New York: Columbia University Press.

———. 2006. *The Attention Revolution: Unlocking the Power of the Focused Mind.* Somerville, MA: Wisdom Publications.

———. 2007. *Contemplative Science: Where Buddhism and Neuroscience Converge.* New York: Columbia University Press.

———. 2009. *Mind in the Balance: Meditation in Science, Buddhism, and Christianity.* New York: Columbia University Press.

———. 2012. *Meditations of a Buddhist Skeptic: A Manifesto for the Mind Sciences and Contemplative Practice.* New York: Columbia University Press.

Wallace, R. K. 1970. "Physiological Effects of Transcendental Meditation." *Science* 167: 1751–54.

Walsh, R. 2007. *The World of Shamanism: New Views of an Ancient Tradition.* Woodbury, MN: Llewellyn.

Wangyal, T. 2002. *Healing with Form, Energy and Light: The Five Elements in Tibetan Shamanism, Tantra, and Dzogchen.* Ithaca, NY: Snow Lion.

Watson, J. B. 1913. "Psychology as the Behaviorist Views It." *Psychological Review* 20: 158–77.

Watson, J. D. 1968. *The Double Helix: A Personal Account of the Discovery of the Structure of DNA.* New York: Atheneum.

West, L. J. (Editor). 1962. *Hallucinations.* New York: Grune & Stratton.

Wheeler, J. A., & Zurek, W.H. (Editors). 1983. *Quantum Theory and Measurement.* Princeton, NJ: Princeton University Press.

Yilmaz, G., Ungan, P., Sebik, O., Uginčius, P., and Türker, K. S. 2014. "Interference of Tonic Muscle Activity on the EEG: A Single Motor Unit Study." *Frontiers in Human Neuroscience* 8: 504. doi: 10.3389/fnhum.2014.00504

ACKNOWLEDGMENTS

Deep gratitude to Geshe Tenzin Wangyal Rinpoche for inviting this book's contributors to Ligmincha to discuss the ideas developed herein. And to the scientists from the Division of Perceptual Studies at the University of Virginia who contributed their writing and their decades of research expertise: Bruce Greyson, Ed Kelly, Emily Kelly, and Jim Tucker. Esteemed teachers, one and all.

Special thanks to colleagues who offered critical readings of substantial portions of the text: Bryce Johnson, Donagh Coleman, Gary Bravo, Gustavo Rodrigues Rocha, Michael Grosso, and Richard Ivry. And thanks for the same to several anonymous reviewers from Columbia University Press. Thanks to colleagues Jacob Dalton, Polly Turner, and Robert Sharf for informative discussions. Very special thanks to my wonderful wife, Kristi Panik, who provided extensive critical commentary and many stylistic changes, greatly improving clarity of presentation throughout.

Special thanks to Alan Wallace for connecting me with Wendy Lochner at Columbia University Press, and to Wendy Lochner and her editorial team for their enthusiastic support of this project.

And deep gratitude to Michael Murphy and Jeff Kripal for truly visionary inspiration; and to Esalen Institute and its Center for Theory and Research for fostering a collegial community of expertise for exploration at the frontiers of consciousness science.

Author royalties from this project will be divided between Ligmincha International and the University of Virginia Division of Perceptual Studies, and used to further the educational and research endeavors of these institutions.

CONTRIBUTORS

Bruce Greyson is Carlson Professor Emeritus of Psychiatry and Neurobehavioral Sciences, and Former Director of the Division of Perceptual Studies (2002–2014), Department of Psychiatry and Neurobehavioral Sciences, University of Virginia School of Medicine, Charlottesville. He is a Distinguished Life Fellow of the American Psychiatric Association, a founder and past president of the International Association for Near-Death Studies, and for twenty-six years was Editor of the *Journal of Near-Death Studies*. He graduated from Cornell University, received his medical doctorate from the State University of New York Upstate Medical College, and completed his psychiatric residency at the University of Virginia. He held faculty appointments in psychiatry at the University of Michigan and the University of Connecticut, where he was Clinical Chief of Psychiatry, before returning to the University of Virginia, where he has practiced and taught psychiatry and carried out research since 1995. His research for the last four decades has focused on near-death experiences and has resulted in more than eighty presentations to national scientific conferences, more than 130 publications in academic medical and psychological journals, and multiple research grants and awards. He is the co-author of *Irreducible Mind: Toward a Psychology for the 21st Century* (2007), and co-editor of *The Near-Death Experience: Problems, Prospects, Perspectives* (1984) and *The Handbook of Near-Death Experiences: Thirty Years of Investigation* (2009).

Edward F. Kelly is professor in the Division of Perceptual Studies, Department of Psychiatry and Neurobehavioral Sciences, University of Virginia School of Medicine, Charlottesville. He received his undergraduate degree in psychology from Yale University, followed by a PhD in psycholinguistics and cognitive science from Harvard in 1971. He then spent fifteen-plus years working nearly full-time in parapsychology, initially at J. B. Rhine's Institute for Parapsychology in Durham, North Carolina, then for ten years through the Department of Electrical Engineering at Duke University, and finally through Spring Creek Institute, a non-profit research institute located in Chapel Hill, North Carolina. During this period he published numerous papers on a variety of experimental, methodological, and theoretical topics in parapsychology and a book on *Computer Recognition of English Word Senses* (with P. J. Stone, 1975). Between 1988 and 2002 he worked with a large neuroscience group at University of North Carolina at Chapel Hill, conducting electroencephalograph (EEG) and functional magnetic resonance imaging (fMRI) studies of human somatosensory cortical adaptation to natural tactile stimuli. He returned full-time to psychical research in 2002, serving as lead author of *Irreducible Mind: Toward a Psychology for the 21st Century* (2007), which systematically updates and reinforces the classic work of F. W. H. Myers in light of the subsequent century of relevant empirical and theoretical effort on mind-brain issues, and as principal editor of *Beyond Physicalism: Toward Reconciliation of Science and Spirituality* (2015). He has now returned to his central long-term research interest—application of modern functional neuroimaging methods to detailed psychophysiological studies of altered states of consciousness and psi in exceptional subjects, and is co-director of the Westphal Neuroimaging Laboratory at the University of Virginia.

Emily Williams Kelly is assistant professor of research in the Division of Perceptual Studies at the University of Virginia, where she has worked since 1978. She received her undergraduate degree from Duke University; an MA in religious studies from the University of Virginia, where her thesis focused on cases of the reincarnation type among the Druses of Lebanon; and a PhD in psychology from the University of Edinburgh, where her dissertation focused on the development of scientific psychology in the late nineteenth century, with an emphasis on F. W. H. Myers and his contributions to psychology. At the University of Virginia, she has conducted research on cases of the reincarnation type, near-death experiences, mediumship, and spontaneous experiences suggestive of survival after death such as apparitions and deathbed visions. She is co-author of *Irreducible Mind: Toward a Psychology for the 21st Century* (2007), and editor of *Science, Survival, and the Self: Selected Writings of Ian Stevenson, M.D.* (2013).

CONTRIBUTORS   199

David E. Presti is teaching professor of Neurobiology, Psychology, and Cognitive Science at the University of California in Berkeley, where he has taught since 1991. Between 1990 and 2000, he also worked as a clinical psychologist in the treatment of addiction and of post-traumatic-stress disorder (PTSD) at the Department of Veterans Affairs Medical Center in San Francisco. He has a master's degree in physics and a doctorate in molecular biology and biophysics from the California Institute of Technology, and a doctorate in clinical psychology from the University of Oregon. Since 2004, he has been teaching neuroscience to Tibetan monastics in India and Bhutan, as part of "Science for Monks & Nuns," a program of monastic science education initiated by the Dalai Lama in 2000. He is author of *Foundational Concepts in Neuroscience: A Brain-Mind Odyssey* (2016).

Jim B. Tucker is Bonner-Lowry Professor of Psychiatry and Neurobehavioral Sciences at the University of Virginia. He is the Director of the Division of Perceptual Studies, where he is continuing the work of Ian Stevenson with children who report memories of previous lives. He is the author of two books about the phenomenon that together have been translated into twenty languages, *Life Before Life: A Scientific Investigation of Children's Memories of Previous Lives* (2005) and *Return to Life: Extraordinary Cases of Children Who Remember Past Lives* (2013), a *New York Times* bestseller. He received a BA in psychology and an MD from the University of North Carolina-Chapel Hill. A board-certified child psychiatrist, he served as medical director of the University of Virginia Child and Family Psychiatry Clinic for nine years.

Geshe Tenzin Wangyal Rinpoche is a master teacher in the Tibetan Bön tradition. Born in India of Tibetan parents, he received monastic education under the guidance of Tibetan Bön and Buddhist masters. In 1991 he was awarded a Rockefeller Fellowship and came to the United States to work at Rice University. In 1992, he founded the Ligmincha Institute in Charlottesville, Virginia. He is a highly respected teacher, known for his ability to make the ancient Tibetan teachings accessible and relevant to the lives of contemporary Westerners. He is the author of ten books: *The Tibetan Yogas of Dream and Sleep* (1998), *Wonders of the Natural Mind: The Essence of Dzogchen in the Native Bon Tradition of Tibet* (2000), *Healing with Form, Energy and Light: the Five Elements in Tibetan Shamanism, Tantra, and Dzogchen* (2002), *Unbounded Wholeness: Dzogchen, Bon and the Logic of the Nonconceptual* (co-authored with Anne Klein) (2006), *Tibetan Sound Healing: Seven Guided*

Practices for Clearing Obstacles, Accessing Positive Qualities and Uncovering Your Inherent Wisdom (2006), *Awakening the Sacred Body* (2011), *Tibetan Yogas of Body, Speech and Mind* (2011), *Awakening the Luminous Mind: Tibetan Meditation for Inner Peace and Joy* (2012), *The True Source of Healing: How the Ancient Tibetan Practice of Soul Retrieval Can Transform and Enrich Your Life* (2015), and *Spontaneous Creativity: Meditations for Manifesting Your Positive Qualities* (2018).

INDEX